Lecture Notes in Biomathematics

Managing Editor: S. Levin

31

Ethan Akin

The Geometry of
Population Genetics

Springer-Verlag
Berlin Heidelberg New York 1979

Author

Ethan Akin
Mathematics Department
The City College
137 Street and Convent Avenue
New York City, NY 10031
USA

Library of Congress Cataloging in Publication Data
Akin, Ethan, 1946-
The geometry of population genetics.
(Lecture notes in biomathematics; 31)
Bibliography: p.
Includes index.
1. Population genetics--Mathematical models. I. Title. II. Series.
QH455.A38   575.1   79-26084

AMS Subject Classifications (1970): 92 A 10, 34 C 05, 34 D 20,
70 G 05, 62 J 10

ISBN-13: 978-3-540-09711-2   e-ISBN-13: 978-3-642-93128-4
DOI: 10.1007/978-3-642-93128-4

Softcover reprint of the hardcover 1st edition 1979

Table of Contents

Introduction

The differential equations which model the action of selection and recombination are nonlinear equations which are impossible to solve explicitly. It is even difficult to describe in general the qualitative behavior of solutions. Recently, Shahshahani began using differential geometry to study these equations [28]. With this monograph I hope to show that his ideas illuminate many aspects of population genetics. Among these are his proof and clarification of Fisher's Fundamental Theorem of Natural Selection and Kimura's Maximum Principle and also the effect of recombination on entropy. We also discover the relationship between two classic measures of genetic distance: the χ^2 measure and the arc-cosine measure.

There are two large applications. The first is a precise definition of the biological concept of degree of epistasis which applies to general (i.e. frequency dependent) forms of selection. The second is the unexpected appearance of cycling. We show that cycles can occur in the two-locus-two-allele model of selection plus recombination even when the fitness numbers are constant (i.e. no frequency dependence).

This work is addressed to two different kinds of readers which accounts for its mode of organization.

For the biologist, Chapter I contains a description of the entire work with brief indications of a proof for the harder results. I imagine a reader with some familiarity with linear algebra and systems of differential equations. Ideal background is Hirsch and Smale's text [15]. In Section 3 we introduce what manifold theory is necessary together with a review of the underlying linear algebra and calculus.

The remaining Chapters are more demanding though the epistasis examples and discussion of position effects in Chapter III are worth a look.

For the mathematician, the technical Chapters II and IV are the heart of the work with Chapter I serving as an introduction and biological orientation. However, some acquaintance with the rudiments of genetics is needed. I recommend "An Introduction to Genetics" by Sturtevant and Beadle (Dover--1962). This is a reprint of a book published in 1939 and so is uncluttered by the fallout of the recent explosive growth of the field.

Here I would like to thank Ms. Kate March for her typing of the manuscript (twice) and the NSF for their support of this work.

I. The Vectorfield Model of Population Genetics

We consider a large population of diploid organisms among whose gametes we distinguish n different types, indexed by a set I. So we describe a member of the population by telling its genotype, a pair ij (= ji) with i and j elements of I. We can describe the population by telling the frequencies of the different genotypes, x_{ij} = the number of organisms with genotype ij. The information in this frequency table is equivalently described by the total population number Σx_{ij} and the distribution of diploid types $\{p_{ij}\}$ where p_{ij} is the fraction of the total population having genotype ij. The diploid zygotes which make up the population are obtained by the pairing of haploid gametes. We will assume that this pairing is random in the Hardy-Weinberg sense. This means that the two gametes in the zygote are independent of one another. It is then sufficient to know the distribution of the haploid gamete types, $\{p_i\}$, and their total number, which we will denote by $|x|$, because $p_{ij} = 2p_i p_j$ ($i \neq j$), $p_{ii} = p_i^2$ and $\Sigma x_{ij} = |x|/2$. If we let R^I denote the n-dimensional vector space of real valued functions on I then the gamete distribution is a vector p in the simplex $\Delta = \{p \in R^I : p_i \geq 0 \text{ and } \Sigma p_i = 1\}$

The genes of the gametes occur on the chromosomes. At each of ℓ different positions, or loci, on the chromosomes are the genes which in the zygote will determine its biological characteristics. For the α position ($\alpha = 1, \ldots \ell$) the n_α different possible genes which can occur constitute a finite set I_α. Thus, I_α is the set of alleles at the α locus. A haploid genotype i is a list of ℓ choices $i_\alpha \in I_\alpha$ for $\alpha = 1, \ldots, \ell$. So the set I of genotypes is the Cartesian product $I = \prod_{\alpha=1}^{\ell} I_\alpha$. The number of genotypes n is the product $\Pi_{\alpha=1}^{\ell} n_\alpha$.

Now let R_α denote the space R^{I_α} and Δ_α be the corresponding simplex. The gamete distribution $p \in \Delta$, is a probability distribution on the product I. It induces a distribution $p^\alpha \in \Delta_\alpha$, namely the marginal distribution on the factor I_α. If $k \in I_\alpha$ then p_k^α (also written $p^\alpha(k)$) is the probability that a random gamete has gene k at the α locus. The map $E^\alpha(p) = p^\alpha$ from Δ to Δ^α is the restriction of the linear mapping $E^\alpha: R^I \to R_\alpha$ defined by

$$(0.1) \quad E^\alpha(x)(k) = \sum \{x(i): \text{ for all } i \text{ with } i_\alpha = k\} \qquad (k \in I_\alpha).$$

Note that we use x_i and $x(i)$ interchangeably for notational conven- ience.

This just means that the probability that k occurs at the α locus is the sum of probabilities p_i where the sum is taken over all genotypes with $i_\alpha = k$.

More generally, if S is any subset of the set of loci $L \equiv \{1,\ldots,\ell\}$, let I_S be the product of the factors I_α for α in S. So $I_S = \Pi_{\alpha \in S} I_\alpha$ is the collection of partial genotypes obtained by ignoring all but the loci in S. For $i \in I$ let i_S denote the projec- tion of i to I_S. So $(i_{S\alpha} = i_\alpha$ for all $\alpha \in S$. Define $R_S = R^{I_S}$ and Δ_S to be the corresponding simplex. p induces a distribution $p^S \equiv E^S(p)$ on the subproduct I_S. $E^S: \Delta \to \Delta_S$ is the restriction of the linear map $E^S: R^I \to R^S$ defined by:

$$(0.2) \quad E^S(x)(k) = \sum \{x(i): \text{ for all } i \text{ with } i_S = k\} \qquad (k \in I_S).$$

So $p^S(k)$ is the probability that the allele k_α occurs at locus α for all of the loci α in S.

If T is another subset of L, disjoint from S, then for $i_S \in I_S$ and $j_T \in I_T$ we denote by $i_S j_T$ the element of $I_{S \cup T}$ whose value at locus α agrees with $(i_S)_\alpha$ for $\alpha \in S$ and with $(j_T)_\alpha$ for $\alpha \in T$. In particular, if we denote by $\widetilde{S}$ the complement of the set S in L then, for $i \in I$, $i = i_S i_{\widetilde{S}}$.

These notations are just bookkeeping devices to keep from writing genotypes and partial genotypes as lists of genes. We turn now to the substance of the model.

1.   The Equations of Selection, Recombination and Mutation.

In the vectorfield or differential equation model of population genetics, evolution is regarded as due to the sum of the effects of selection, recombination and mutation. Assuming the Hardy-Weinberg condition, we represent each of these by a vectorfield on the space of gametic genotype distributions, Δ.

We have assumed that the diploid genotype of a member of the population determines it biological characteristics, among these are two rates: a reproductive rate and a death rate. Each zygote of type ij is assumed to have an average of $b_{ij} dt$ offspring in a time interval of length dt and to have probability $d_{ij} dt$ of dying in the same time interval. By an offspring of a zygote we mean two gametes given to newborns which are zygotes receiving complementary gametes from other members of the population. On average the two gametes contributed will be an i and a j. Since we are only counting gametes we can think of an offspring of an ij zygote as a gain of an ij zygote.

The gain or loss of an ij zygote causes the gain or loss of one (if i ≠ j) or two (if i = j) i gametes. Thus, if we define fitness $m_{ij} = b_{ij} - d_{ij}$ the change in the number of i gametes in time dt is given by:

$$dx_i = (2m_{ii}x_{ii} + \sum_{j \neq i} m_{ij}x_{ij})dt = x_i m_i dt.$$

Here we define $m_i = \Sigma_j\, m_{ij}p_j$ and get the last equation from the Hardy-Weinberg assumption in the form $x_{ij} = 2p_i p_j \cdot (|x|/2) = x_i p_j\, (i \neq j)$ and $2x_{ii} = 2p_i^2(|x|/2) = x_i p_i$, where $x_i = p_i \cdot (|x|)$ is the number of gametes of type i.

So we get the first selection equation:

(1.1)
$$\frac{dx_i}{dt} = x_i m_i.$$

Recall that $|x| = \Sigma\, x_i$ is the total number of gametes. So:

(1.2)
$$\frac{d|x|}{dt} = |x|\bar{m}.$$

Since $p_i = x_i/|x|$ the quotient rule implies that:

(1.3)
$$\frac{dp_i}{dt} = \frac{1}{|x|}(\frac{dx_i}{dt} - p_i \frac{d|x|}{dt}).$$

Applying this to (1.1) and (1.2) we get:

(1.4)
$$\frac{dp_i}{dt} = p_i(m_i - \bar{m}).$$

Here $\bar{m} = \Sigma_i p_i m_i = \Sigma_{i,j} p_i p_j m_{ij}$ is mean fitness. Note that we write b_{ij}, m_{ij}, etc. as functions of unordered pairs or, as in the latter equation, as symmetric functions of ordered pairs.

Recall that the offspring of an ij zygote consisted of i and j gametes. This assumes that there is no recombination. The recombination term in the equation is the correction which must be included if there is.

Let S be a subset of $L = \{1,\ldots,\ell\}$ the set of loci. With probability r^S an ij zygote will suffer a series of crossovers so that i and j will exchange genetic material exactly in the loci of S, or equivalently, exactly in the loci of the complement, $\widetilde{S} = L - S$. The offspring will then consists of $\bar{i} = i_S j_{\widetilde{S}}$ and $\bar{j} = j_S i_{\widetilde{S}}$ gametes, where $i_S j_{\widetilde{S}}$ is the element of I agreeing with i at the loci of S and with j at the loci of $\widetilde{S}$. The recombination probabilities themselves can be under genetic control in which case we write r^S_{ij} for the probability of an S-exchange in a parent of type ij. r^S and r^S_{ij} really depend only on the pair $\{S,\widetilde{S}\}$ and so we will assume $r^S_{ij} = r^{\widetilde{S}}_{ij} =$ one half of the actual recombination probability.

In the most important example the loci are arranged in order on a single chromosome. When a single crossover between the μ and $\mu + 1$ loci $(1 \leq \mu < \ell)$ occurs then $S = S_\mu \equiv \{\alpha \in L: \alpha \leq \mu\}$.

We saw above that $b_{ij}dt$ times $|x|p_i p_j$ gametes of type i are contributed to the gene pool as offspring of the ij zygotes in a time interval of length dt. Of these the fraction r^S_{ij} are lost by S-recombination. On the other hand, $r^S_{\bar{i}\bar{j}} b_{\bar{i}\bar{j}} |x| p_{\bar{i}} p_{\bar{j}} dt$ gametes of type i are contributed by S-recombination of the $\bar{i}\bar{j}$ zygotes. So the term which must be added to equation (1.1) to correct for recombination is:

(1.5) $$\left(\frac{dx_i}{dt}\right)_R = -|x| \sum_{\bar{j},S} r^S_{ij} b_{ij} p_i p_j - r^S_{\bar{i}\bar{j}} b_{\bar{i}\bar{j}} p_{\bar{i}} p_{\bar{j}}.$$

If we sum these terms on i we get zero, meaning that the effect of the correction on the gamete population growth rate, $(d|x|/dt)_R$, is zero. So the correction term for dp_i/dt is given by (see (1.3)):

$$(1.6) \qquad (\frac{dp_i}{dt})_R = -\sum_{j,S} r^S_{ij} b_{ij} p_i p_j - r^S_{\bar{i}\bar{j}} b_{\bar{i}\bar{j}} p_{\bar{i}} p_{\bar{j}}$$

The form of the recombination term is simpler if we assume that r^S_{ij} and b_{ij} are <u>completely</u> <u>symmetric</u> meaning $r^S_{ij} = r^S_{\bar{i}\bar{j}}$ and $b_{ij} = b_{\bar{i}\bar{j}}$ for all i,j and S. That r^S_{ij}, b_{ij} and d_{ij} are symmetric in i and j, eg. $b_{ij} = b_{ji}$, is just a result of thinking of the genotype of the zygote as an unordered pair of gametes. The complete symmetry assumption means that the phenotypic characteristics of the zygote, namely b_{ij}, d_{ij} and the r^S_{ij}'s depend only on the genes and not on how they are associated on the chromosomes. For example, in the two locus, two allele case this means that the "coupling" and "repulsion" hetero-zygotes have the same phenotype. The failure of complete symmetry is one form of what geneticists refer to as <u>position</u> <u>effects</u>.

If complete symmetry holds then we can rewrite equation (1.6):

$$(1.7) \qquad (\frac{dp_i}{dt})_R = -\sum_{j,S} r^S_{ij} b_{ij} (p_i p_j - p_{\bar{i}} p_{\bar{j}}).$$

Be careful here of useful but misleading notation. $\bar{i}$ and $\bar{j}$ each depend on i,j and S.

The final member of our trinity is the correction due to muta-tion. We take the equation straight from Wright [35, p. 369].

Let n_{ij} be the relative rate by which i gametes are trans-formed to j gametes by mutation when $i \neq j$. Define $n_{i*} = \Sigma n_{ij}$, summing on all $j \neq i$. The correction for (1.1) due to mutation is denoted $(dx_i/dt)_N$. It is given by:

$$(1.8) \qquad (\frac{dx_i}{dt})_N = \sum_{j \neq i} x_j n_{ji} - x_i n_{i*} = |x| \sum_{j \neq i} p_j n_{ji} - p_i n_{i*}.$$

This says that the net rate of change of x_i is the difference between the absolute rates at which i gametes are produced and lost. Again the sum on i is zero and so $(d|x|/dt)_N$ equals zero. So:

$$(1.9) \qquad (\frac{dp_i}{dt})_N = \sum_{j \neq i} p_j n_{ji} - p_i n_{i*}.$$

If we assume that mutations occur independently at the separate loci then the n_{ij}'s have a special form which we will look at later.

These equations are all in the text books of population genetics, eg. Crow and Kimura [6], although the notation which makes recombination tractable for multilocus models is essentially due to Shahshahani [28].

I won't say much about the biological simplification built into the model. For example, the assumption that the phenotype is determined by the genotype means that we ignore or average out environmental effects. Also the model has no age structure as we lump all the zygotes together and don't include any lag time for development. These matters are better described by biologists. Jacquard [17], for example, has a particularly careful discussion of the role of random mating and large population size in such models. However, there are two points of interest which are really in the mathematical domain.

Postulating the Hardy-Weinberg condition is a mathematically odd way to proceed. What one ought to do is start with a model for zygotic frequencies and then prove that the Hardy-Weinberg condition follows. That is, show that under certain conditions every solution of the zygotic differential equation tends toward the region (the submanifold, actually) where the Hardy-Weinberg condition holds, or at least that any solution which begins in the Hardy-Weinberg region remains there. I didn't do it because it doesn't work. Hoppensteadt has looked at such a model [16, Sec. II.2]. Only if the death rates d_{ij} are constant (i.e. independent of the genotype ij) is the Hardy-Weinberg set preserved. He shows, however, that if the d_{ij}'s are nearly constant then there is an invariant submanifold close to the Hardy-Weinberg submanifold. This is one reason, among several, that the model is limited to the case of "slow selection".

The other point has to do with the number of loci to which the model is applicable. One of the central ideas of this paper is that the introduction by Shahshahani of differential geometric methods to the study of these classical equations should allow us to get beyond the small models of the two locus two allele case in studying the interaction between selection and recombination. But the vectorfield model is still only a medium-sized model. While it is designed to get beyond the two-locus models there is still a certain size limitation. Once the number of genotypes n gets to be the order of magnitude of the population size or greater, it no longer makes sense to think of the gene pool as a continuous flow of genotype frequencies because each genotype will appear in the pool only a small whole number of times. This is the truism of genetic uniqueness. If there are n_a alleles at every locus then $n = (n_a)^{\ell}$. So we must really assume

(1.10)
$$\ell \ln n_a < \ln|x|.$$

Since $\ln 20 < 3$, if we are dealing with 20 alleles per locus and a population of 1,000,000 or so then ℓ can't be much bigger than 3 or 4. If there are only 2 alleles per locus then the model is reasonable for 15 or 16 loci. In any case the vectorfield model can only deal with a tiny number of loci compared to the actual genome of most species.

2. Multivariate Analysis and Types of Epistasis.

Consider a metric character ξ_i or ξ_{ij} which we think of as a real-valued function of the gametic or zygotic genotype. In the realm of genetic statistics we fix the gamete probability distribution p_i and regard these functions as random variables on the set of genotypes. So the usual statistical functions are defined such as the <u>mean</u>:

$$(2.1) \qquad \bar{\xi} = \sum_i p_i \xi_i \quad \text{or} \quad \sum_{i,j} p_i p_j \xi_{ij}$$

and the <u>variance</u>:

$$(2.2) \qquad \text{Var}(\xi) = \sum_i p_i (\xi_i - \bar{\xi})^2 \quad \text{or} \quad \sum_{i,j} p_i p_j (\xi_{ij} - \bar{\xi})^2.$$

Given two such random variables ξ and η we define their <u>covariance</u>:

$$(2.3) \qquad \text{Cov}(\xi, \eta) = \sum_i p_i (\xi_i - \bar{\xi})(\eta_i - \bar{\eta}) \quad \text{or} \quad \sum_{i,j} p_i p_j (\xi_{ij} - \bar{\xi})(\eta_{ij} - \bar{\eta}).$$

The historical bridge between the genetic statistics of a fixed population and the evolution problem is in the response of various metric traits to artificial selection. It becomes important to determine the contribution of different loci or blocs of loci to the total effect as well as the interaction between the loci. For example, a character is called <u>additive</u> if the total effect is the sum of effects at the various loci. This means that the function ξ_i from I to R can be written as a sum $\Sigma_\alpha \, \varphi_{i_\alpha}^\alpha$ where φ^α is a function on the alleles I_α at the α locus. A positive character is called <u>multiplicative</u> if its log is additive. In the case where ξ_i is gametic fitness, m_i, additivity is also referred to as the absence of epistasis or <u>zero epistasis</u>. We will use the term <u>epistasis</u> to refer to interaction between the loci for any character under consideration. We formalize different types of epistasis.

Let K be a nonempty collection of subsets of L, the set of loci, such that $S_1 \in K$ and $S_2 \subset S_1$ imply $S_2 \in K$. We will call such a collection a <u>complex</u> <u>of</u> <u>loci</u> or <u>gene</u> <u>complex</u>. If K is a complex of loci then we will say that a character ξ, regarded as a function $\xi: I \to R$, is <u>carried</u> <u>by</u> K or has K-<u>type</u> <u>epistasis</u> if there exist functions $\varphi^S: I_S \to R$ for $S \in K$ such that

(2.4)
$$\xi_i = \sum \{\varphi^S(i_S): S \in K\}.$$

So a function ξ has K-type epistasis if it is the sum of functions each depending only on a bloc of loci in K. For example, $L^{(0)}$ consisting of the empty set and each single locus (i.e. $L^{(0)} = \{\emptyset, \{1\}, \{2\}, \ldots, \{\ell\}\}$) is a complex called the <u>zero-skeleton</u> of L. $L^{(0)}$ type epistasis is just what we called zero-epistasis above. Note that a function $\varphi^\emptyset$ depending on none of the loci is just a constant. Similarly, ξ depends only on pairs of loci, or <u>one-dimensional epistasis</u>, is associated with the complex $L^{(1)}$ consisting of the sets in $L^{(0)}$ and all pairs of loci. In general, for $s \leq \ell-1$ we can define the s-<u>skeleton</u> $L^{(s)} = \{S \subset L: S$ consists of $s+1$ or fewer loci$\}$. We will refer to $L^{(s)}$ type epistasis as s-<u>dimensional</u> <u>epistasis</u>. The geneticist would say that such a character exhibits $(s+1)$-way interactions.

If K_1 and K_2 are complexes then the union, written $K_1 \vee K_2$, and the intersection, written $K_1 \wedge K_2$, are again complexes. If S is any bloc of loci (i.e. $S \subset L$) then S together with all of its subsets is a complex which we will also refer to as S. One reason for this deliberate ambiguity is that if $S_1 \subset S$ and $\varphi^{S_1}: I_{S_1} \to R$ then we can regard φ^{S_1} as a function on I_S by $\varphi^{S_1}(k) \equiv \varphi^{S_1}(k_{S_1})$ for $k \in I_S$. Here k_{S_1} is the projection to the subproduct I_{S_1} which just forgets the part of the genotype not in S_1. So if $\varphi^S: I_S \to R$ we can regard $\varphi^S + \varphi^{S_1}$ as a function on I_S. Thus, we can amalgamate together the functions on subsets of S to get just one function on I_S. Doing this in formula (2.4) we see that ξ has S-type epistasis if it can be written as a single function of i_S. This means that ξ depends only on the loci in S. That is, variation of the genotype in the loci not in S has no effect on the value of the character ξ. This suggests a generalization of zero-epistasis different from s-dimensional epistasis. Suppose $\{T_a: a = 1,\ldots,\ell'\}$ is a set of pairwise disjoint subsets of L, i.e. each locus occurs in at most one set T_a. Regarding each T_a as a complex we can form the union, as complexes, and so get the <u>disjoint</u> <u>bloc</u> <u>model</u> $T_1 \vee \ldots \vee T_{\ell'}$. A character shows this kind of epistasis if it is the sum of effects each depending only on the loci in one of the blocs T_a, i.e. it is additive between the blocs. $L^{(0)}$ is a disjoint bloc model where the T_a's each consist of a single locus.

One remark about language. A geneticist would use the term gene complex to refer to a collection of associated loci, in other words, to what I am calling a bloc. Mathematically, these blocs are the simplices of the complex K.

In studying epistasis it is important to have a test to see whether a character ξ satisfies K-type epistasis. For example, when $K = L^{(0)}$ we are given a function $\xi(i) = \xi(i_1, \ldots, i_\ell)$ which we can think of as a function in ℓ different variables and we want to know when ξ can be written as a sum:

(2.5)
$$\xi(i_1, \ldots, i_\ell) = \varphi^1(i_1) + \varphi^2(i_2) + \ldots + \varphi^\ell(i_\ell)$$

The variable i_α is discrete as it varies over the finite set I_α. However, the answer to the question is easier when the variables i_α are continuous real variables. Consider the case when $\ell = 2$ and so ξ is a function of (i_1, i_2) with i_1 and i_2 elements of R. Suppose that ξ is smooth meaning that all partial derivatives are defined and continuous. Clearly, if $\xi(i_1, i_2) = \varphi^1(i_1) + \varphi^2(i_2)$ then the mixed partial derivative:

(2.6)
$$\frac{\partial^2 \xi}{\partial i_1 \partial i_2} = 0.$$

Conversely, if (2.6) holds then $\frac{\partial \xi}{\partial i_2}$ doesn't depend on i_1 and neither does its integral with respect to i_2 which we will call $\varphi^2(i_2)$. ξ and φ^2 have the same partial derivative with respect to i_2 and so:

$$\frac{\partial(\xi - \varphi^2)}{\partial i_2} = 0.$$

Thus, $\xi - \varphi^2$ doesn't depend on i_2 and so is a function $\varphi^1(i_1)$. This proves (2.5) from (2.6) in the case $\ell = 2$. A similar argument using mathematical induction on ℓ proves that (2.5) holds if and only if

$$(2.7) \qquad \frac{\partial^2 \xi}{\partial i_\alpha \partial i_\beta} = 0 \qquad \text{for all} \quad \alpha \neq \beta \in L.$$

In general, for smooth functions with ℓ real variables the analogue of K-type epistasis corresponds to the vanishing of various mixed partial derivatives. For example $L^{(s)}$-type epistasis corresponds to the vanishing of all s + 2-mixed partials.

In the discrete variable case we will derive general formulae for detecting K-type epistasis in Chapter II. The basic tool in constructing the formulae is the discrete analogue of the partial derivative operator.

So far we have made no use of the probability distribution p which weighs the points of I. It is used in the analysis of the variance of ξ.

Suppose that ξ is a character which does show some epistasis. We can ask: what is the best zero-epistasis approximation ξ^0 to ξ? This means first, that ξ^0 has zero-epistasis and, second, that the mean of ξ^0 equals the mean of ξ. The mean comes in because the mean of ξ is the best approximation of ξ by a constant. Third, the variance of the "error" $\xi - \xi^0$ is assumed to be smaller for the choice ξ^0 than for any other choice of approximator satisfying the first two conditions. So we are using a least-squares notion of approximation. As we will see in the next section this sort of approximation arises naturally in linear algebra and from such general considerations it follows that a best approximation always exists. It also follows that the variance of ξ is the sum of the variances of ξ^0 and of the error $\xi - \xi^0$. So we can answer the question: how much of the variance of ξ can be attributed to interaction between the loci? The answer is the variance of $\xi - \xi^0$. If most of the variance of ξ lies in ξ^0 then we can throw away ξ and use the approximation ξ^0 instead and thus suppose that the character is additive. How good the approximation has to be depends on the tolerances of the application at hand. If too much of the variance remains in the error, we can look to pairwise interactions and take the best $L^{(1)}$ approximation of $\xi - \xi^0$ which we call ξ^1. Then $\xi^0 + \xi^1$ is the best approximation of ξ having only $L^{(1)}$ type epistasis. Continuing by approximating $\xi - \xi^0 - \xi^1$ among $L^{(2)}$ functions we get ξ^2 and so forth. The details of this partitioning of the variance of ξ into terms involving higher and higher interactions is a standard device in genetic statistics (see for example Kempthorne [20 Chaps. 13 and 19]). It would clearly be useful to have a general formula for the best K-type epistasis approximation to ξ. In an important special case such a formula can be derived using the discrete partial derivative operators mentioned above. We carry this out in Chapter II. The special case is when the loci are in linkage equilibrium meaning that the different loci are probabilistically independent. Equivalently, the distribution p on the product set I is just the product distribution obtained from the marginal distributions p^α on the factors I_α. This is equivalent to the formula:

$$(2.8) \qquad p_i = p(i_1, \ldots, i_\ell) = p^1(i_1) \cdot p^2(i_2) \cdot \ldots \cdot p^\ell(i_\ell).$$

The set of distributions in linkage equilibrium is a subset Λ of the set Δ of all distributions. Shahshahani calls Λ the <u>Wright mani-fold</u> and we will meet it again. For now notice that if we think of p as a metric trait, it is after all a real-valued function on i, then if p is in Λ, (2.8) implies that p is multiplicative, i.e. the log, $\ell n\ p_i$, has zero-epistasis. The converse is true and we will see later than this partly accounts for the key role of Λ in the mathematics.

The projections of ξ to its approximations are not so nice if the distribution is not in Λ. This is one reason why the text-books tend to assume linkage equilibrium.

3. Euclidean Vector Spaces and Riemannian Manifolds.

Suppose that f is a function from R^{n_1} to R^{n_2}, a list of n_2 real functions of n_1 real variables. More generally, suppose that f is a function between vector spaces V_1 and V_2, a vector-valued function of a vector variable. What then does differentiation mean? What is the derivative of f at a point x of the domain? Recall from a first course in calculus that for a real function of a real variable ($n_1 = n_2 = 1$) the derivative, f'(x), is a number. This mis-leads one from the general answer: The derivative of f at a point x is a function, but a linear function. It is the linear mapping which is in some sense (quite different from a least squares idea) the best approximation to f near x by a linear map. Looked at this way, the purpose of calculus is to convert problems about non-linear functions to problems about linear ones (see Palais [27, Chap. 1]). In short, calculus is generalized linear algebra. So before discussing manifolds, which are places where one can do calculus, we first review some ideas from linear algebra.

A real vector space or linear space is a set whose elements are called vectors together with a definition of addition of vectors and of multiplication of vectors by real numbers (also called scalars). Addition and multiplication are required to satisfy certain standard axioms. The most important example is R^n, the set of ordered n-tuples of real numbers with coordinate-wise addition and multiplication.

$$(x_1, \ldots, x_n) + (y_1, \ldots, y_n) = (x_1 + y_1, \ldots, x_n + y_n)$$

(3.1)

$$t(x_1, \ldots, x_n) = (tx_1, \ldots, tx_n).$$

Most of the examples we will meet are subspaces of some R^n. A subset of a vector space is a subspace, i.e. is a vector space in its own right, if it is closed under addition and scalar multiplication. For the three dimensional space R^3 the subspaces, other than the trivial extremes of R^3 itself and the set consisting of 0 alone, are the lines and planes which contain 0. Notice that a line or plane which does not contain 0 is not a subspace. It is neither closed under addition nor under scalar multiplication.

The axiomatic viewpoint is important even with these examples because it is used to construct new vector spaces. For example, the set of linear maps between two vector spaces is itself a vector space. A linear map $T: V_1 \to V_2$ is a function which relates the vector space operations:

$$T(\xi + \eta) = T(\xi) + T(\eta) \qquad \xi, \eta \in V_1$$

(3.2)

$$T(t \cdot \xi) = t \cdot T(\xi) \qquad \xi \in V_1 \text{ and } t \in R.$$

Here the operations on the left are occurring in V_1 and those on the right are in V_2. These linearity properties are very special. For example, the false assumption of linearity underlies many mistakes in elementary algebra, eg. $\sqrt{x+y} = \sqrt{x} + \sqrt{y}$ (false). The set of all linear maps between V_1 and V_2, denoted $L(V_1, V_2)$, becomes a vector space when we define addition and scalar multiplication by:

$$(T_1 + T_2)(\xi) = T_1(\xi) + T_2(\xi) \qquad (T_1, T_2 \in L(V_1, V_2), \; \xi \in V_1)$$

(3.3)

$$(t \cdot T)(\xi) = t(T(\xi)) \qquad (T \in L(V_1, V_2), \xi \in V_1, \; t \in R)$$

Here the operations on the right are in V_2 and are defining the linear maps $T_1 + T_2$ and $t \cdot T$ by describing their value on a typical element ξ of V_1. It is a good exercise to show that $T_1 + T_2$ and $t \cdot T$ so defined are linear maps, i.e. they satisfy (3.2).

Actually, this definition of addition and multiplication for functions comes directly from (3.1). We can regard an n-tuple $(x_1, \ldots, x_n)$ as a function $x: \{1, \ldots, n\} \to R$ with $x(i) = x_i$. In general, if I is any set and R^I is the set of all functions from I to R we define:

$$(x + y)(i) = x(i) + y(i) \qquad x, y \in R^I, \; i \in I$$

(3.4)

$$(t \cdot x)(i) = t \cdot x(i) \qquad x \in R^I, \; i \in I, \; t \in R.$$

When I is the set $\{1, \ldots, n\}$ this definition coincides with (3.1).

The most important space of linear maps is the dual space of a vector space V also called the space of _linear_ _forms_ on V. The dual space, denoted V*, is L(V,R). It is the space of linear maps from V to the reals. If $\xi \in V$ and $\omega \in V*$ then the value of ω at ξ, i.e. $\omega(\xi)$, is also denoted $\langle\omega,\xi\rangle$ and is then called the _Kronecker_ _product_. So

$$(3.5) \qquad \langle\omega,\xi\rangle = \omega(\xi) \qquad \omega \in V*, \ \xi \in V.$$

Regarded as a function of two variables, ω and ξ, the product $\langle \ , \ \rangle$ is bilinear, that is, it is linear in each variable alone with the other held fixed.

The linear operations allow us to construct new vectors from old. If $\xi^1,\ldots,\xi^n$ is a list of vectors in V and $x_1,\ldots,x_n$ is a list of scalars then the vector $\xi = x_1\xi^1 +\ldots+ x_n\xi^n = \Sigma_i \ x_i\xi^i$ is called the linear combination of the vectors $\xi^1,\ldots,\xi^n$ with coefficients $x_1,\ldots,x_n$. The list of vectors is called linearly independent if we can equate coefficients, that is, if $\Sigma \ x_i\xi^i = \Sigma \ y_i\xi^i$ implies $x_i = y_i$ for $i = 1,\ldots,n$. So we can translate a vector equation into a list of scalar equations. If $\xi^3 = \xi^1 + \xi^2$, for example, then $\{\xi^1,\xi^2,\xi^3\}$ is not linearly independent, and so is called linearly dependent, because $1\cdot\xi^1 + 1\cdot\xi^2 + (-1)\cdot\xi^3 = 0\cdot\xi^1 + 0\cdot\xi^2 + 0\cdot\xi^3$. If every vector in V is a linear combination of the linearly indepen-dent set $\{\xi^1,\ldots,\xi^n\}$, so that the list also spans V, then we say that the set is a basis for V. The structure theorem of linear algebra says that while a vector space has in general many different bases, any two bases have the same number of elements. This number is called the dimension of V. If $\{\xi^1,\ldots,\xi^n\}$ is any list of vectors in V then there is a linear map T: $R^n \to V$ defined by:

(3.6)
$$T(x_1,\ldots,x_n) = \sum x_i \xi^i.$$

If $\{\xi^1,\ldots,\xi^n\}$ is a basis then this map is onto, meaning that every

vector in V is in the image of T, because the set spans. It is

one-to-one, meaning that no two lists of coefficients hit the same

vector in V, because the set is linearly independent. A one-to-one

and onto linear map T: $V_1 \to V_2$ is called a linear isomorphism. If

T is a linear isomorphism then the inverse map T^{-1}: $V_2 \to V_1$ is de-

fined and is also linear. For the special case defined by (3.6) with

$\{\xi^1,\ldots,\xi^n\}$ a basis for V this inverse map associates to every

vector ξ in V the list of coefficients $(x_1,\ldots,x_n)$ such that

$\xi = \Sigma x_i \xi^i$. The scalars $(x_1,\ldots,x_n)$ are then called the <u>coordinates</u>

of ξ with respect to the basis. A different basis will in general

give different lists of coordinates for the vector ξ. In general, if

the vectors of a basis are indexed by a set I, then the above con-

struction gives an isomorphism of R^I with V. On R^I itself the

<u>standard basis</u> $\{e^i: i \in I\}$ is defined by letting e^i be zero at all

points of I except i at which e_i is one. So if we define the

Kronecker delta δ_{ij} by $\delta_{ij} = 0$ if $i \neq j$ and $\delta_{ii} = 1$ then the basis is

defined by:

(3.7)
$$e^i(j) = \delta_{ij}.$$

We call it the standard basis because the coordinate map $R^I \to R^I$ is

the identity, i.e. $x = \Sigma x(i)e^i$.

These coordinate maps show that all vector spaces having a

finite basis, the so-called finite dimensional vector spaces, are just copies under some isomorphism of R^n where n is the dimension

of the space.

If $\{\xi^1,\ldots,\xi^n\}$ is a basis chosen for V_1 and $\{\eta^1,\ldots,\eta^m\}$ is a

basis chosen for V_2 then this choice of bases associates to every

linear map $T: V_1 \to V_2$ an $m \times n$ matrix (a_{ij}) (i = 1,...,m and

j = 1,...,n) satisfying the equation

(3.8)
$$y_i = \sum_j a_{ij}x_j \qquad i = 1,\ldots,m$$

whenever $(x_1,\ldots,x_n)$ are the coordinates of ξ in V_1 with respect to

$\{\xi^1,\ldots,\xi^n\}$ and $(y_1,\ldots,y_m)$ are the coordinates of $T(\xi)$ in V_2 with

respect to $\{\eta^1,\ldots,\eta^m\}$. a_{ij} is defined by letting the jth column

a_{ij}, i = 1,...,m be the coordinates of $T(\xi^j)$ with respect to the η

basis. The composition of two linear maps $T_2 \cdot T_1: V_1 \to V_3$, where

$T_1: V_1 \to V_2$ and $T_2: V_2 \to V_3$, is again a linear map and the matrix of

the composition $T_2 \cdot T_1$ is the product in the same order of the two

matrices for T_2 and T_1. In fact, this is the reason behind the odd

definition of the product of matrices.

In addition to the algebraic concepts of addition and multi-

plication we need a definition of the distance between vectors before

the limit operation in calculus or any concept of approximation makes

sense. We make the definition by using a _Euclidean_ _metric_ or inner

product on a vector space V. This is a function (,): $V \times V \to R$,

i.e. a real-valued function of two vector variables. (,) is

bilinear and symmetric (i.e. $(\xi,\eta) = (\eta,\xi)$). Furthermore, it

satisfies:

(3.9)
$$(\xi,\xi) > 0 \quad \text{if} \quad \xi \in V \quad \text{and} \quad \xi \neq 0.$$

Notice that, because of bilinearity, $(0,\xi) = (\xi,0) = 0$ for any vector ξ in V.

This allows us to define the length, or norm, or absolute value, of a vector by

$$(3.10) \qquad \|\xi\| = (\xi,\xi)^{1/2}.$$

By analogy with the real numbers we define the distance between ξ and η to be the length of the difference, $\|\xi-\eta\| = \|\eta-\xi\|$.

On the space R^I there is the so-called usual inner product:

$$(3.11) \qquad (\xi,\eta) = \sum_i \xi_i \eta_i.$$

More generally, if p is a distribution on I, i.e. $p \in \Delta$, then we can define the covariance metric:

$$(3.12) \qquad _p(\xi,\eta) = \sum_i p_i \xi_i \eta_i.$$

$_p(\ ,\ )$ is symmetric and bilinear but satisfies (3.9) only if p is an interior distribution, meaning that $p_i > 0$ for all i. So for (3.12) to define a Euclidean metric we must have $p \in \overset{\circ}{\Delta} = \{p \in R^I : \Sigma_i p_i = 1$ and $p_i > 0$ for all $i \in I\}$.

The inner product gives more than just the length. For any inner product $(\ ,\ )$ on a vector space V:

$$(3.13) \qquad (\xi,\eta) = \|\xi\| \cdot \|\eta\| \cdot \cos\theta$$

where θ is the angle between the two vectors. In R^2 or R^3 with the usual inner product this is a theorem of trigonometry (the law of cosines). For a general vector space equipped with a fixed

Euclidean metric--we will call such a space a <u>Euclidean vector space</u>--
(3.13) is used to define the angle θ. Then by using bilinearity to
expand $\|\xi \pm \eta\|^2 = (\xi \pm \eta, \xi \pm \eta)$ we get the law of cosines as a theorem:

(3.14)
$$\|\xi \pm \eta\|^2 = \|\xi\|^2 + \|\eta\|^2 \pm 2(\xi, \eta)$$
$$= \|\xi\|^2 + \|\eta\|^2 \pm 2\|\xi\|\|\eta\|\cos \theta.$$

It is a theorem that $(\xi, \eta)/\|\xi\| \cdot \|\eta\|$ always has absolute value
at most 1 (Schwarz inequality) and so it makes sense to regard it as
the cosine of an angle. In particular, this angle is a right angle
if and only if the cosine is zero. So ξ and η are perpendicular,
or orthogonal, if and only if $(\xi, \eta) = 0$. With respect to the usual
inner product on R^I distinct members of the standard basis are orthogonal. Furthermore, the length of each basis vector is 1. We can
summarize this by saying that for $\xi^i = e^i$ (i ∈ I):

(3.15)
$$(\xi^i, \xi^j) = \delta_{ij} \qquad i, j \in I.$$

In general, in a Euclidean vector space a basis which satisfies (3.14)
is called an orthonormal basis. For example, with respect to $_p(\, , \,)$
the basis $\{\xi^i = p_i^{-1/2} e^i\}$ (i ∈ I) is orthonormal. A general procedure
called Gram-Schmidt orthogonalization process constructs an orthonormal basis starting from any basis.

If the basis of V is orthonormal then the linear isomorphism
from R^I to V defined by the basis and equation (3.6) is also an
isometry with the usual metric on R^I and the given metric on V. A
linear map T: $V_1 \to V_2$ between Euclidean vector spaces is called an
isometry if it preserves the metrics:

(3.16) $(\xi,\eta)_1 = (T(\xi),T(\eta))_2$ $\xi,\eta \in V_1$.

An isometry preserves length and distance and so is one-to-one. It is an isomorphism if it is onto. In that case the inverse map is also an isometry.

Since an orthonormal basis always exists we see that every finite-dimensional Euclidean vector space is isometrically isomorphic to R^n with the usual metric where n is the dimension of the space.

Every linear map T: $R \to V$ can be naturally identified with a vector ξ in V, namely, $\xi = T(1)$ because $T(t) = tT(1) = t\xi$. This gives a linear isomorphism between L(R,V) and the space V itself.

Using the inner product we can get a--quite different--isomorphism between V and the space of linear maps from V to R, i.e. the dual space V*. Every vector $\eta \in V$ defines a linear form $\eta*: V \to R$ via the inner product, namely $\eta*(\xi) = (\eta,\xi)$. The association of $\eta*$ with η defines a linear map of V into V* by bilinearity of (,). It is easily seen to be one-to-one because if $\eta* = 0$ then $\eta*(\eta) = (\eta,\eta) = 0$ and so $\eta = 0$. The Riesz representation theorem says that this map is onto and so defines a linear isomorphism between V and its dual:

1 Theorem: Let V be a finite dimensional Euclidean space. For every linear form $\omega: V \to R$ there exists a unique vector $\eta \in V$ such that $\omega(\xi) = (\eta,\xi)$ for all $\xi \in V$.

Proof: Choose an orthonormal basis $\{\xi^i\}$ for V. With respect to this basis, and the number 1 chosen as a basis for R, ω is represented by a $1 \times n$ matrix. These n numbers are the coordinates of η with respect to the ξ-basis. In more detail, if the matrix is (a_i)

then $\omega(\xi) = \Sigma\ a_i x_i$ where $\{x_i\}$ are the coordinates of ξ with respect to $\{\xi^i\}$, i.e. $\xi = \Sigma\ x_j \xi^j$. Define $\eta = \Sigma\ a_i \xi^i$. Then by (3.15) $(\eta, \xi) = \Sigma\ a_i x_j (\xi^i, \xi^j) = \Sigma\ a_i x_j \delta_{ij} = \Sigma\ a_i x_i = \omega(\xi)$. So this η works, i.e. $\eta^* = \omega$. It is the only one which does because the map $\eta \to \eta^*$ is one-to-one.

QED

This simple result has many profound consequences. For the moment, we will use it to define the least squares approximations which we used in the previous section.

<u>2 Theorem</u>: Let V be a finite dimensional Euclidean space and A be a subspace of V. If ξ is a vector in V there is a unique vector ξ_A in A satisfying the following equivalent conditions:

    (i) $(\eta, \xi) = (\eta, \xi_A)$ for all η in A.

    (ii) $\xi - \xi_A$ is orthogonal to every vector η in A.

    (iii) For every vector η in A the Pythogorean identity holds:

(3.17)
$$\|\xi - \eta\|^2 = \|\xi - \xi_A\|^2 + \|\xi_A - \eta\|^2.$$

In particular, if $\eta = 0$ then we have:

(3.18)
$$\|\xi\|^2 = \|\xi - \xi_A\|^2 + \|\xi_A\|^2.$$

<u>Proof:</u> ξ^* is the linear form on V defined by $\xi^*(\eta) = (\eta, \xi)$. Restricting to A we get a linear form on A and so by Thm. 1 there is a unique vector ξ_A in A such that $\xi^*(\eta) = \xi_A^*(\eta)$ for all η in A. This proves that a unique ξ_A satisfying (i) exists. Since $(\eta, \xi) = (\eta, \xi_A)$ if and only if $(\eta, \xi - \xi_A) = 0$, (i) is equivalent to (ii). (ii) implies (iii) by the law of cosines (3.14) applied to $\xi - \xi_A$, $\xi_A - \eta$ and their sum. Conversely, if (iii) holds we can

replace η by the vector $\eta + \xi_A$ in (3.17) and get

$$\| (\xi - \xi_A) - \eta \|^2 = \| \xi - \xi_A \|^2 + \| \eta \|^2.$$

Applying (3.14) to $\xi - \xi_A$, η and their difference we get
$(\xi - \xi_A, \eta) = 0$. So (iii) implies (ii).   QED

Define the function $P_A : V \to A$ by $P_A(\xi) = \xi_A$. Using (i) it is easy to check that P_A is a linear map and that $P_A(\xi) = \xi$ if ξ lay in A to begin with. P_A is called the _orthogonal projection_ of V onto A.

Equation (3.17) explains the sense in which ξ_A is the best approximation of ξ by a vector in A.

When $V = R^I$ with the usual inner product ξ_A is the usual least squares approximation. When the inner product is $_p(\, , \,)$ with $p \in \dot{\Delta}$, $\| \xi \|^2$ the variance of ξ plus the square of the mean of ξ. So (3.18) justifies our remarks about partitioning variance in Sec. 2.

Now we describe the elements of advanced calculus as they appear in modern texts like Edwards [8] or Spivak [32].

Suppose U is an open subset of a Euclidean vector space V_1. This means that whenever a point $x \in U$ then all points sufficiently close to x also lie in U, i.e. there exists $\epsilon > 0$ depending on x such that $\| h \| < \epsilon$ implies $x + h \in U$. Let f be a function on U with values in a Euclidean vector space V_2. So $f : U \to V_2$. The derivative of f at a point $x \in U$ is a linear map written $d_x f : V_1 \to V_2$. It is the unique linear map such that the function $f(x) + d_x f(h)$ (with x fixed and h varying) gives the best approximation to f near x, i.e. to $f(x + h)$. This means that not only does the error term $= f(x + h) - f(x) - d_x f(h)$ approach 0 as h approaches 0 (and so x + h approaches x), but the ratio between the error term and the length of h also goes to zero. We write this as follows:

(3.19) $$f(x + h) = f(x) + d_x f(h) + o(h)$$

where the error term denoted o(h) is defined for $\|h\|_1$ sufficiently small and satisfies:

(3.20) $$\|o(h)\|_2 / \|h\|_1 \longrightarrow 0 \qquad \text{as} \qquad \|h\|_1 \longrightarrow 0.$$

We will usually drop the subscripts on the length which here are reminders of which Euclidean metric (whether in V_1 or V_2) is being used.

The derivative of a function need not exist, for example, $f(x) = x^{1/3}$ defined from R to R is not differentiable at x = 0, but unless otherwise mentioned all of the functions we will look at are _smooth_ or C^∞ meaning that all derivatives exist and are continuous.

When $V_1 = R^n$ and $V_2 = R^m$ then with respect to the standard bases the derivative $d_x f$ can be represented by an m × n matrix. This matrix is just the Jacobian matrix of partial derivatives. If $f(x) = (f_1(x), \ldots, f_m(x))$ and $x = (x_1, \ldots, x_n)$ then the matrix a_{ij} is given by $a_{ij} = \partial f_i / \partial x_j$ (i = 1, ..., m and j = 1, ..., n).

Taking the derivative itself as a linear operation. If $f, g: U \to V_2$ and $t \in R$ then $d_x(tf + g) = t(d_x f) + (d_x g)$. So in the standard case the Jacobian matrix of the sum of two functions is the sum of the corresponding Jacobian matrices. We will also need the chain rule which says that the derivative of a composite map is the composition of the derivatives. If $f: V_1 \to V_2$ and $g: V_2 \to V_3$ then the composite function is $g \cdot f \colon V_1 \to V_3$ defined by $g \cdot f(x) = g(f(x))$

for $x \in V_1$. Now for x in V_1 we can take the derivatives of f and

$g \cdot f$ at x and the derivative of g at $f(x)$. We get linear maps

$d_x f \colon V_1 \to V_2$, $d_{f(x)} g \colon V_2 \to V_3$ and $d_x (g \cdot f) \colon V_1 \to V_3$. The chain rule

says:

(3.21) $$d_x (g \cdot f) = (d_{f(x)} g) \cdot (d_x f).$$

In the standard case this implies that the Jacobian of the composite

$g \cdot f$ is the product of the Jacobians of g and of f.

When $V_1 = R$ and $V_2 = V$ so that $f(t)$ is a vector-valued function

of a real variable, $d_t f$ is a linear map from R to V. We saw

earlier that such a map can be identified with the vector $d_t f(1)$ and

we denote this vector $f'(t)$. So $d_t f(s) = s f'(t)$. $f'(t)$ is the limit

of the familiar difference quotient $(f(t + s) - f(t))/s$ as s

approaches 0. On the other hand, when $V_1 = V$ and $V_2 = R$, $d_x f$ is a

linear form on V called the differential of f at x. If $f \colon U \to R$

then the differential $df \colon U \to V^*$ associates to x, the form $d_x f$.

Now if we use the Euclidean metric on V, the Riesz representation

theorem (Thm. 1) associates to $d_x f$ a vector in V. This is the

gradient of f at x denoted $\mathrm{grad}_x f$. It is defined by:

(3.22) $$d_x f(h) = (\mathrm{grad}_x f, h).$$

The gradient depends on the particular Euclidean metric on V. Up

to now we have only needed the metric to make the limit statements

like (3.20) make sense. But any Euclidean metric will give the same

idea of limit, the same topology, on V. So the derivatives like $d_x f$

are independent of the choice of metric. This is not true of the gradient and we will later see different kinds of gradients.

If h is a vector of unit length in V, then $d_xf(h)$ is called
the directional derivative of f in the direction h. It is the
limit of the difference quotient $(f(x + sh) - f(x))/s$ as s
approaches 0. By (3.22) and (3.13), $d_xf(h) = \|grad_xf\|\cos \theta$ where θ
is the angle between h and the gradient. Clearly, this is largest
when $\cos \theta = 1$, i.e. $\theta = 0$. So the gradient has the direction of
greatest increase of the function f.

In general, the derivative d_xf describes the behavior of f
near x. So calculus is used to solve local problems. For example,
if f: R → R then $f'(t) = 0$ and $f''(t) < 0$ implies f has a local
maximum at t, i.e. $f(t) > f(s)$ for s different from but close to
t. It may happen that far from t f becomes larger than $f(t)$.

The most important example of a problem which can be solved
locally by calculus is described by the inverse function theorem.

Suppose U_1 is open in V_1 and $f: U → V_2$. f is called a
diffeomorphism if it has a smooth inverse map, i.e. if f maps U_1
one-to-one and onto an open set U_2 and the inverse map $f^{-1}: U_2 → V_1$
is smooth. When f is a diffeomorphism and $x \in U_1$ then the chain
rule implies that the linear map d_xf is a linear isomorphism and that
its inverse is the derivative of f^{-1} taken at $f(x)$. Thus, if f is
invertible so is its derivative at each point. The inverse function
theorem is the converse, at least locally. For the proof see
[8, p. 185] or [32, p. 35].

3 Theorem: Let $f: U → V_2$ be a smooth map with U open in V_1 and let
$x \in U$. If the derivative $d_xf: V_1 → V_2$ is a linear isomorphism then
f is locally a diffeomorphism near x, i.e. there exists an open set

$U_1 \subset U$ with $x \in U_1$ and f restricted to U_1 is a diffeomorphism.

If the set I has n elements and k is a whole number with $k \leq n$, then a k-dimensional manifold in the vector space R^I is a subset M of R^I which looks locally, near each point, like a curved piece of a k-dimensional subspace. There are two equivalent ways of making this precise.

First, we can define M near $x \in M$ explicitly by defining a coordinate system on M near x. This is a function h: $U \rightarrow M$ where U is open in R^k and h maps U one-to-one and onto all of the points of M near x (i.e. the intersection of M with some open set in R^I). h is assumed to have rank k. This means that if we regard h as a function from U to R^I the derivative $d_u h: R^k \rightarrow R^I$ is one-to-one at every point u of U. This description is called explicit because it parametrizes the points of M near x by k real parameters. For example, the piece of the circle of radius 1 in the interior of the first quadrant of R^2 is the image of the function $f(t) = (\cos t, \sin t)$ with t varying in the open interval of R between 0 and $\pi/2$. Similar pieces can be constructed near any point of the circle. This example illustrates the typical fact that often no coordinate system can be found which works on the entire manifold. The manifold is obtained by gluing together many coordinate patches.

The implicit description of the manifold near x is as the level surface of a family of functions. This means we have a function $F: G \rightarrow R^{n-k}$, with G some open subset of R^I containing x, such that the points of the manifold in G are precisely the solutions of the

equations $F(y) = \xi$ for some fixed vector ξ in R^{n-k}, i.e.
$M \cap G = F^{-1}(\{\xi\})$. F is assumed to have rank $n - k$ at all points of
$M \cap G$. This means that the derivative $d_y F: R^I \to R^{n-k}$ is onto for
every point y in $M \cap G$. We can think of F as a list of $n - k$
scalar functions and the equation $F(y) = \xi$ as a l st of $n - k$ constraints, which reduce the number of degrees of freedom
(= dimension) from n to k. Frequently an implicit description can be given for the entire manifold. For example, the $n - 1$
dimensional sphere of radius r in R^I is given by the single scalar
equation $F(y) = r^2$ where $F(y) = \Sigma y_i^2$. The subset $\overset{\circ}{\Delta}$ of R^I is defined
by the equation $F(y) = 1$ where $F(y) = \Sigma y_i$. Here the open set G
consists of the set of vectors with positive coordinates.

Just as the derivative at a point of a function is a linear
approximation to the function, there is at every point of a manifold
a linear subspace which approximates the manifold.

A path through x in M is a function v from an open interval in R to M such that $v(t) = x$ for some t in the interval.
Taking the derivative at t we get the vector $v'(t)$ which is called
a tangent vector at x. The collection of all tangent vectors at x
is a linear subspace of R^I are called the <u>tangent space</u> of M at x
and denoted $T_x M$. It is not clear from this definition that $T_x M$ is a
subspace, but $T_x M$ can also be defined using the explicit or implicit
description of M near x. If $h: U \to M$ with U open in R^k is a
coordinate system near x then every path in M through x can be
described using these coordinates. It then follows from the chain
rule (3.22) that $T_x M$ is the image of the linear map $d_x h$. Since $d_x h$
is one-to-one $T_x M$ is a k dimensional subspace of R^I. On the other
hand, if $F: G \to R^{n-k}$ with $M \cap G = F^{-1}(\{\xi\})$ then every path in M maps

under F to a constant path in R^{n-k}. Since constants have derivative

0, T_xM is the kernel of the linear map $d_xF: R^I \to R^{n-k}$, i.e.

$T_xM = \{y \in R^I: d_xF(y) = 0\}$. For example, if $F(y) = \Sigma_i y_i$ then

$d_xF(y) = \Sigma_i y_i$ and so the tangent space $T_p\overset{\circ}{\Delta} = \{y \in R^I: \Sigma_i y_i = 0\}$ for

all p in $\overset{\circ}{\Delta}$. On the other hand, if $F(y) = \Sigma_i y_i^2$ then

$d_xF(y) = 2 \Sigma_i x_i y_i = 2(x,y)$ where $(\ ,\ )$ is the usual inner product.

So for the sphere of radius r, the tangent space at x consists of

all vectors orthogonal to x. Notice that $T_p\overset{\circ}{\Delta}$ is the same subspace

for all p, but the tangent space of the sphere at x changes as x

changes.

If M_1 is a manifold in R^{I_1}, M_2 is a manifold in R^{I_2} and f is

a function from M_1 to M_2, we can extend the definition of f to a

function from U to R^{I_2} where U is some open set in R^{I_1} containing

M_1. Then for x in M_1 we can define the derivative $d_xf: R^{I_1} \to R^{I_2}$.

There are many different ways of extending f and d_xf will depend

on which extension is used. However d_xf maps T_xM_1 into $T_{f(x)}M_2$ and

this part of d_xf does not depend on the choice of extension, so we can

define the linear map $d_xf: T_xM_1 \to T_{f(x)}M_2$ without ambiguity. The

reason is that if v is a path in M_1 through x then the composition

$f \cdot v$ is a path in M_2 through $f(x)$, and $(f \cdot v)'(t) = d_xf(v'(t))$ by the

chain rule. This allows us to do calculus on manifolds. For example,

if d_xf is a linear isomorphism of T_xM_1 onto $T_{f(x)}M_2$ then one can

extend the inverse function theorem to show that f is a diffeo-

morphism between some open set of M_1 containing x and some open set

of M_2 containing $f(x)$.

In particular, if $R = M_2 = R^{I_2}$ then the differential of f, df

associates to each $x \in M_1$ the linear form d_xf on T_xM_1.

Dual to the idea of the differential of a function is the idea

of a vectorfield. A vectorfield X on an open set U of R^I is just
a function X: U → R^I. A vectorfield on a manifold M in R^I is a
function X: M → R^I such that $X(x) \in T_xM$ for all $x \in M$, i.e. X is
always tangent to M. Via the Kronecker product, eg. (3.5), we can
associate to a function f: M → R and a vectorfield X on M a new
function <df,X> defined by:

(3.23) $$\langle df,X \rangle (x) = \langle d_xf, X(x) \rangle = d_xf(X(x)) \qquad x \in M.$$

With f fixed we can regard (3.23) as a way that functions operate
on vectorfields to get new functions, or with X fixed we can regard
(3.23) as the way a vectorfield operates on functions. From the
latter viewpoint we define the vectorfield ∂_i on R^I to be constantly
the standard basis vector e^i. The notation comes from the fact that

$$\langle df, \partial_i \rangle = \frac{\partial f}{\partial x_i} \quad \text{at each point} \quad x \quad \text{of } R^I.$$

This is because $\langle df, \partial_i \rangle$ is just the directional derivative in the e^i
direction. Since $\{e^i\}$ is a basis every vectorfield X on M can
be written uniquely as a linear combination $X = \Sigma\, X_i \partial_i$ where each X_i
is a real-valued function on M. Note that the ∂_i's themselves
usually do not lie in T_xM and so not every choice of function X_i will
define a vectorfield on M.

The Kronecker product is bilinear and so

(3.24) $$\langle df,X \rangle = \sum X_i \langle df, \partial_i \rangle = \sum X_i \frac{\partial f}{\partial x_i}.$$

Note here that as the ∂_i's are not vectorfields on M the expressions $\partial f/\partial x_i$ will depend on the choice of extension of f to a

neighborhood of M. However, $\langle df, X \rangle$ itself does not depend on this choice.

A vectorfield on M is the manifold analogue of a differential equation. A solution path for the vectorfield is a path $v(t)$ in M such that for all t:

$$(3.25) \qquad\qquad v'(t) = X(v(t)).$$

In local coordinates it is easy to check that this is just an ordinary differential equation in the k coordinates.

An important example of a vectorfield, a gradient field, requires the notion of a Riemannian metric.

Any inner product on R^I restricts to an inner product on each subspace and in particular on the tangent spaces of a manifold M. However, in many applications including those of this paper the inner product which arises naturally from the problem will be different at different points. A <u>Riemannian</u> <u>metric</u> on a manifold M is a smooth choice of inner product $(\, , \,)_x$ for each subspace $T_x M$. A manifold equipped with a Riemannian metric is called a Riemannian manifold.

On a Riemannian manifold M we define the gradient ∇f of a function $f: M \to R$. $d_x f$ is a linear form on $T_x M$ and so by Thm. 1 there exists a unique vector $\nabla_x f \in T_x M$ such that:

$$(3.26) \qquad (\nabla_x f, X)_x = \langle d_x f, X \rangle = d_x f(X) \qquad x \in M, \quad X \in T_x M.$$

The vectorfield ∇f thus depends not only on $f: M \to R$ but also on the Riemannian metric. This is in contrast to df which depends only on f.

If M_1 is a submanifold of M, i.e. another manifold in R^I with $M_1 \subset M$, then for $x \in M_1$ the tangent space $T_x M_1$ is a subspace of $T_x M$

since every path in M_1 lies in M. So a Riemannian metric on M restricts to define one on M_1. If f: M → R then f restricts to a function $f|M_1$: M_1 → R. Now for x ∈ M_1 the two gradients $\nabla_x f$ and $\nabla_x(f|M_1)$ both satisfy (3.26) for vectors X ∈ T_xM_1. In addition, $\nabla_x(f|M_1)$ itself lies in T_xM_1. So by Thm. 2 $\nabla_x(f|M_1)$ is the perpendicular projection of $\nabla_x f$ into T_xM_1. For a connected submanifold, i.e. every pair of points in M_1 can be joined by a path in M_1, this implies the following:

3 Proposition: Let M_1 be a connected submanifold of a manifold M and let f: M → R. The following conditions are equivalent:

    (i)   f is constant on M_1.

    (ii) $d_x f(X) = 0$ for all X ∈ T_xM_1, x ∈ M_1.

    (iii) $\nabla_x(f|M_1) = 0$ for all x ∈ M_1.

    (iv) $\nabla_x f$ in T_xM is orthogonal to the subspace T_xM_1 for all
        x ∈ M_1.

On a Riemannian manifold the arclength of a path is defined. Thinking of t as time the velocity vector at time t of the path v(t) in M is $v'(t) ∈ T_{v(t)}M$. The speed is the length of the velocity measured using the inner product at v(t). So $\|v'(t)\| = (v'(t)\ v'(t))^{1/2}_{v(t)}$. Integrating the speed we get the length of the path. This enables us to define the distance between two points x_1 and x_2 of M as the greatest lower bound of the lengths of all paths in M connecting x_1 and x_2. A path which achieves this length--the shortest distance between the two points--is called a geodesic. For example, on the sphere with the Riemannian metric obtained by restricting the usual inner product on R^I, geodesics are pieces of great circles, i.e. the intersection of the sphere with two dimensional subspaces (planes through zero).

A diffeomorphism between two Riemannian manifolds M_1 and M_2 is called an <u>isometry</u> if $d_x g: T_x M_1 \to T_{g(x)} M_2$ is an isometry at every point x of M_1. While diffeomorphisms between pieces of manifolds of the same dimension are quite common, isometries are quite rare. This is related to the rigidity of geometry as opposed to the flexibility of topology ("rubber-sheet geometry"). Isometries preserve the structures of Riemannian geometry. They map geodesics to geodesics and relate gradient vectorfields, i.e. $d_x g(\nabla_x (f \cdot g)) = \nabla_{g(x)} f$ if g is an isometry but not, in general, if it is not.

4.   The Shahshahani Metric

Fisher's Fundamental Theorem of Natural Selection says that along the solutions curves of the selection differential equation, (1.4), mean fitness, $\bar{m}$, is constantly increasing. Kimura's Maximum Principle says that the direction of motion is the direction of greatest increase. These results suggest that the selection vectorfield on $\overset{\bullet}{\Lambda}$ associated with the selection differential equation (sensu (3.25)), should be the gradient of $\bar{m}$. When one computes the gradient one gets the wrong equation. However, we saw in the previous section that the concept of the gradient of a function depends upon a choice of Riemannian metric. When we compute grad $\bar{m}$ we are using the metric obtained from the inner product on R^I. A careful statement of Kimura's theorem, eg. see [6, p. 230] shows that the concept of direction, i.e. unit vector, means unit variance. Since the definition of variance depends on the distribution p it becomes clear that we should look for a non-constant Riemannian metric on $\overset{\bullet}{\Lambda}$. The

appropriate Riemannian metric was discovered by Shahshahani in [28].
As we will see this metric plays a central role in interpreting selection, recombination and mutation geometrically.

For I a set of n elements define the subsets of R^I:
$P = \{x: x_i \geq 0 \text{ for all } i \geq 0\}$ and $\overset{\circ}{P} = \{x: x_i > 0 \text{ for all } i\}$. Then
$\Delta = \{p \in P: \Sigma p_i = (p,1) = 1\}$ and $\overset{\circ}{\Delta} = \Delta \cap \overset{\circ}{P} = \{p \in \overset{\circ}{P}: \Sigma p_i = (p,1) = 1\}$.
Here $(,)$ is the usual inner product on R^I and 1 is the function
on I constantly 1, regarded as a vector in R^I. $\overset{\circ}{P}$, as an open subset of R^I, is a manifold with tangent space R^I at every point. Recall
that $\overset{\circ}{\Delta}$ is a manifold of dimension $n - 1$ whose tangent space at every
point is the subspace $(R^I)_0 = \{X: \Sigma X_i = (X,1) = 0\}$. For $x \in \overset{\circ}{P}$ we
define the <u>Shahshahani metric</u> $(,)_x$ on $R^I = T_x\overset{\circ}{P}$ by:

$$(4.1) \qquad (X,Y)_x = \sum_i x_i^{-1} X_i Y_i \qquad X,Y \in R^I, \ x \in \overset{\circ}{P}.$$

In particular, regarding x itself as a vector in R^I we have

$$(4.2) \qquad (X,x)_x = (X,1) = \sum_i X_i.$$

So for $p \in \overset{\circ}{\Delta}$, $(p,p)_p = 1$ and $T_p\overset{\circ}{\Delta} = \{X \in R^I: (X,p)_p = 0\}$. With respect
to the usual inner product the constant vector $n^{-1/2} 1$ is the unit
vector in R^I perpendicular to $T_p\overset{\circ}{\Delta}$. But with respect to $(,)_p$ it is
p itself which is the perpendicular unit vector.

The Shahshahani metric restricts to a metric on $\overset{\circ}{\Delta}$. Here if
we think of $X \in T_p\overset{\circ}{\Delta} = (R^I)_0$ as a little change in the distribution p,
we can write the square of the magnitude two ways:

$$(4.3) \qquad \|X\|_p^2 = \sum \frac{X_i^2}{p_i} = \sum p_i \left(\frac{X_i}{p_i}\right)^2.$$

In the first form we weight the square of X_i by the inverse of p_i because we regard changes of smaller values of p_i as more significant. In the second form, we regard X_i/p_i as the relative change and average the squares of the relative changes by the distribution.

In the first form this is referred to as the χ^2 measure of genetic distance (see Jacquard, [17, p. 427] or Kempthorne, [20, p. 178]). This suggests an interesting coordinate change which among other applications reveals the relationship between two measures of genetic distance.

1 Theorem: The smooth map $f: R^I \to P$ defined by $f(z) = x$ with $x_i = z_i^2/4$ admits a smooth inverse on $\overset{\circ}{P}$ $g: \overset{\circ}{P} \to \overset{\circ}{P}$ defined by $g(x) = z$ with $z_i = 2\sqrt{x_i}$. g is an isometry between $\overset{\circ}{P}$ with the Shahshahani metric and $\overset{\circ}{P}$ with the usual Euclidean metric. If $S_2 = \{z \in R^I : \Sigma_i z_i^2 = 4\}$ is the sphere of radius 2 then $f^{-1}(\Delta) = S_2$ and so g restricts to an isometry between $\overset{\circ}{\Delta}$ and $S_2 \cap \overset{\circ}{P}$.

Proof: For $x \in \overset{\circ}{P}$ the derivative is given by $d_x g(X) = r$ with $r_i = X_i/\sqrt{x_i}$ and so:

$$(d_x g(X), d_x g(Y)) = \sum_i \left(\frac{X_i}{\sqrt{x_i}}\right)\left(\frac{Y_i}{\sqrt{x_i}}\right) = \sum_i \frac{X_i Y_i}{x_i} = (X,Y)_x. \qquad \text{QED}$$

2 Corollary: If $p,q \in \overset{\circ}{\Delta}$ then the geodesic distance between p and q with respect to the Shahshahani metric on $\overset{\circ}{\Delta}$ is given by

$$d(p,q) = 2 \text{ arc cos } \left(\sum_i p_i q_i\right).$$

Here the principal value of arc cos is chosen, measured between 0 and $\pi/2$ radians. This measure of distance is the arc measure of Cavilla-Sforza and Edwards (see Jacquard, [17, p. 425]).

Proof: An isometry preserves geodesic distance and so the distance between p and q in $\overset{\circ}{\Delta}$ is the same as the distance between g(p) and g(q) in $S \cap \overset{\circ}{P}$. But on the sphere geodesic distance is measured along the arc of great circles. This arc distance is the radius (= 2) times the angle between g(p) and g(q) measured in radians. The cosine of this angle is just the usual inner product between the unit vectors $\frac{1}{2}$ g(p) and $\frac{1}{2}$ g(q). Finally, since these vectors lie in $S \cap \overset{\circ}{P}$ (and so does the arc between them) the inner product is positive and the arc cosine is between 0 and $\pi/2$.   QED

Thus, the arc measure of genetic distance is obtained by integrating the χ^2 metric along a geodesic in $\overset{\circ}{\Delta}$. Alternatively, the χ^2 metric is the "infinitesimal" version of the arc cosine metric.

The second sum in (4.3) has a differential equation intepretation. Let $X = \Sigma \, X_i \partial_i$ be a vectorfield on $\overset{\circ}{P}$ with the X_i's real functions of $x \in \overset{\circ}{P}$. By (3.25) the associated differential equation is dx/dt = X, or in coordinate form:

(4.4) $$\frac{dx_i}{dt} = X_i \qquad (i \in I).$$

Thus, $X_i(x)$ is the absolute growth rate of x_i when the system is at state x. If we define ξ by $X_i = x_i \xi_i$ or $\xi_i = X_i/x_i$, then:

(4.5) $$\frac{dx_i}{dt} = x_i \xi_i \quad \text{or} \quad \frac{d \ln x_i}{dt} = \xi_i \qquad (i \in I).$$

So $\xi_i(x)$ is the relative growth rate of x_i when the system is at state x. Define $|x| = \Sigma \, x_i = (x,1)$ to be the total population size and $p_i = x_i/|x|$ to be the distribution at x. The mean of ξ is $\bar{\xi} = \Sigma \, p_i \xi_i$ and so $|x|\bar{\xi} = \Sigma \, x_i \xi_i$. If we sum the left hand system in

(4.5) we get

(4.6) $$\frac{d|x|}{dt} = |x|\bar{\xi} \quad \text{or} \quad \frac{d \ln |x|}{dt} = \bar{\xi}.$$

Subtracting the logarithmic derivatives we get:

(4.7) $$\frac{dp_i}{dt} = p_i(\xi_i - \bar{\xi}) \quad \text{or} \quad \frac{d \ln p_i}{dt} = \xi_i - \bar{\xi}.$$

So ξ_i normalized to mean zero is the relative growth rate of p_i.
(4.3) suggests the relationship between the Shahshahani metric on
absolute rates and the covariance inner product on relative rates.

3 Proposition: For $x \in \overset{\circ}{P}$, let $|x| = \Sigma\, x_i$ and $p = x/|x| \in \overset{\circ}{\Delta}$. Define
$F_x: R^I \to R^I$ by $F_x(\xi) = X$ with $X_i = x_i\xi_i$. If $F_x(\xi) = X$ and $F_x(\eta) = Y$,
then

(4.8) $$(X,Y)_x = {}_p(\xi,\eta) \cdot |x|$$

where $(\ ,\ )_x$ is the Shahshahani metric on $R^I = T_x\overset{\circ}{P}$ and ${}_p(\xi,\eta)$
$= \Sigma\, p_i\xi_i\eta_i$. In particular, if $x = p$ then F_p is an isometry between
R^I with inner product ${}_p(\ ,\ )$ and $R^I = T_p\overset{\circ}{P}$ with the inner product
$(\ ,\ )_p$. In this case, the mean $\bar{\xi} = {}_p(\xi,1) = (X,p)_p = (X,1)$. So F_p
maps the vectors with zero mean onto the tangent space $T_p\overset{\circ}{\Delta}$. Furthermore, normalizing the vector ξ to mean zero corresponds to project-
ing the vector $X \in T_p\overset{\circ}{P}$ to $T_p\overset{\circ}{\Delta}$ orthogonally with respect to the
Shahshahani metric.

Proof: (4.8) and the mean equations are easy direct computations.
The projection result says

(4.9) $$F_p(\xi - \bar{\xi}) = X - (X,p)_p p = X - (X,1)p$$

which follows from $F_p(1) = p$ and the mean equations. The vector on the right is the orthogonal projection (see Thm. 3.2).   QED

If $f: \overset{\circ}{P} \to R$ is a smooth function then the gradient of f with respect to the usual inner product is:

(4.10)
$$\text{grad } f = \sum \frac{\partial f}{\partial x_i} \partial_i .$$

For if $X = \Sigma\, x_i \partial_i$, $(\text{grad } f, X) = \Sigma \frac{\partial f}{\partial x_i} X_i = d_x f(X)$. For the gradient with respect to the Shahshahani metric denoted $\triangledown f$, we must have $(\triangledown_x f, X)_x = d_x f(X)$ and so:

(4.11)
$$\triangledown_x f = \sum x_i \frac{\partial f}{\partial x_i} \partial_i = F_x(\text{grad}_x f).$$

This means that a vectorfield on $\overset{\circ}{P}$ is the gradient of f with respect to the usual metric if the absolute rate of change for x_i is the partial derivative $\partial f/\partial x_i$. It is the gradient of f with respect to the Shahshahani metric if the relative rate of change for x_i is $\partial f/\partial x_i$.

For the restriction of f to $\overset{\circ}{\Delta}$, Prop. 3.3 says that the gradient of $f|\overset{\circ}{\Delta}$ with respect to $\overset{\circ}{\Delta}$ is the orthogonal projection of the gradient of f with respect to $\overset{\circ}{P}$. Denoting this gradient $\overline{\triangledown} f$ or $\overline{\triangledown}(f|\overset{\circ}{\Delta})$ we have from (4.9):

(4.12)
$$\overline{\triangledown}_p f = \triangledown_p f - (\triangledown_p f, p)_p p = F_p(\text{grad}_p f - \overline{\text{grad}_p f}).$$

We should mention that Shahshahani's definition of his metric differs from ours by a factor of $|x|$. His original definition has the advantage that the selection vectorfield on $\overset{\circ}{P}$ (and not just on $\overset{\circ}{\Delta}$) is a gradient with respect to his metric. This is not true in our case. We have chosen our definition to get Thm. 1 and (4.11). The two definitions agree on $\overset{\bullet}{\Delta}$ where most of our applications will take place.

We conclude this section by defining and computing the gradients of certain special functions. In the table below, $a, b \in R^I$ and $|b| = (b,1) = 0$.

4 Table

$f(x)$	$\nabla_x f$	$\bar{\nabla}_p f \equiv \nabla_p(f\vert\overset{\bullet}{\Delta})$
$E^a(x) = \Sigma \, x_i a_i$	$\Sigma \, x_i a_i \partial_i$	$\Sigma \, p_i(a_i - \bar{a})\partial_i$
$L^b(x) = \Sigma \, b_i \, \ell n \, x_i$	$\Sigma \, b_i \partial_i$	$\Sigma \, b_i \partial_i$
$H(x) = -\Sigma \, x_i \, \ell n \, x_i$	$-\Sigma \, x_i(\ell n \, x_i + 1)\partial_i$	$-\Sigma \, p_i(\ell n \, p_i - H(p))\partial_i$

The following inner product equations will be useful:

$$(\nabla_x E^a, \nabla_x E^{\hat{a}})_x = \Sigma \, x_i a_i \hat{a}_i$$

$$(\bar{\nabla}_p E^a, \bar{\nabla}_p E^{\hat{a}})_p = \mathrm{Cov}_p(a_i, \hat{a}_i),$$

(4.13)

$$(\nabla_x E^a, \nabla_x L^b)_x = \Sigma \, a_i b_i = (a,b)$$

$$(\nabla_x H, \nabla_x L^b)_x = -L^b(x).$$

5. The Product Theorems and Epistasis

Recall the map $E^\alpha: \Delta \to \Delta^\alpha$ which associates to a distribution p on the product I the marginal distribution p^α on the factor I_α (c.f. equation (0.1)). Taking the product we get a map $E = \Pi_\alpha E^\alpha: \Delta \to \Pi_\alpha \Delta^\alpha$ associating to p the list of marginal distributions. E maps

$\mathring{\Delta}$ onto $\Pi_\alpha \mathring{\Delta}^\alpha$ i.e. the marginal distributions of an interior distribution are interior. The dimension of $\Pi_\alpha \mathring{\Delta}^\alpha$ is $\Sigma_\alpha (n_\alpha - 1) = (\Sigma_\alpha n_\alpha) - \ell$ which is usually much smaller than the dimension of $\mathring{\Delta}$ which is $n-1 = (\Pi_\alpha n_\alpha) - 1$. So for any list of marginal distributions there are an infinite number of distributions on the product with the given marginals. In other words, the distributions of the individual genes by no means determine the distribution of genotypes. One needs some additional information describing the linkage of the genes. In the two-locus-two-allele case Δ is 3 dimensional and $\Delta_1 \times \Delta_2$ is 2 dimensional. So one parameter is sufficient to describe the degree of linkage. This parameter is sometimes called the coefficient of linkage disequilibrium. We now describe the generalization of this parameter and its relatives to the multilocus case.

If $i,j \in I$ and S is a subset of $L = \{1,\dots,\ell\}$, we introduced the notation $\bar{i},\bar{j}$ to stand for $\bar{i} = i_S j_{\tilde{S}}$ and $\bar{j} = j_S i_{\tilde{S}}$ ($\tilde{S}$ is the complement $L - S$). The pair of gametes $\bar{i}$ and $\bar{j}$ are obtained from the pair i and j by exchange of genetic material at exactly the loci in the set S. Now define:

(5.1)
$$d_{ij}^S = p_i p_j - p_{\bar{i}} p_{\bar{j}}$$

(5.2)
$$L_{ij}^S = \ln p_i p_j / p_{\bar{i}} p_{\bar{j}} = \ln p_i p_j - \ln p_{\bar{i}} p_{\bar{j}}$$
$$= \ln p_i - \ln p_{\bar{i}} - \ln p_{\bar{j}} + \ln p_j.$$

Note that d_{ij}^S is defined for $p \in \Delta$ but L_{ij}^S is only defined for $p \in \mathring{\Delta}$.

<u>1 Proposition</u>: For any fixed $S \subset L$ and $p \in \Delta$ $d_{ij}^S = 0$ for all i and j in I if and only if the loci of S and the loci of $\tilde{S}$ are independent with respect to p, that is:

(5.3) $\qquad p_i = p^S(i_S) \cdot p^{\tilde{S}}(i_{\tilde{S}})$ for all $i \in I$.

Furthermore, if $p \in \overset{\circ}{\Delta}$ these conditions are equivalent to: $L_{ij}^S = 0$ for all i and j in I.

Proof: $d_{ij}^S = 0$ if and only if $p_i p_j = p_{\bar{i}} p_{\bar{j}}$ and if $p \in \overset{\circ}{\Delta}$ this is true if and only if $\ell n\ p_i p_j = \ell n\ p_{\bar{i}} p_{\bar{j}}$ or $L_{ij}^S = 0$. If (5.3) then $p_i p_j = p^S(i_S) p^S(j_S) p^{\tilde{S}}(i_{\tilde{S}}) p^{\tilde{S}}(j_{\tilde{S}}) = p_{\bar{i}} p_{\bar{j}}$. On the other hand if $p_i p_j = p_{\bar{i}} p_{\bar{j}}$ for all i and j we can sum on j. On the left we get $\Sigma\ p_i p_j = p_i\ \Sigma\ p_j = p_i$. On the right we can think of j as two independent variables j_S in I_S and $j_{\tilde{S}}$ in $I_{\tilde{S}}$. Summing $p_{\bar{i}}$ over the $j_{\tilde{S}}$ variables we get $p^S(i_S)$ by (0.2). Summing $p_{\bar{j}}$ over the j_S variables we get $p^{\tilde{S}}(i_{\tilde{S}})$. Hence (5.3). $\qquad$ QED

2 Corollary: For $p \in \Delta$, $d_{ij}^S = 0$ for all $i, j \in I$ and all $S \subset L$ if and only if p lies in the Wright manifold Λ, i.e. equation (2.8) holds. For $p \in \overset{\circ}{\Delta}$, $L_{ij}^S = 0$ for all $i, j \in I$ and all $S \subset L$ if and only if p lies in $\overset{\circ}{\Lambda} \equiv \Lambda \cap \overset{\circ}{\Delta}$.

Proof: (5.3) holds for all S. This means each locus is independent of the rest with respect to p. So p is a product distribution. QED

Now fix $i, j \in I$ and $S \subset L$ and let p vary over $\overset{\circ}{\Delta}$. L_{ij}^S becomes a real-valued function on $\overset{\circ}{\Delta}$. In the notation of Table 4 of the previous section L_{ij}^S is of the restriction to $\overset{\circ}{\Delta}$ of a function $L^b(x)$ where b_k is 0 except for $k = i, j$ and $k = \bar{i}, \bar{j}$ where it is 1 and -1 respectively. Each of the coordinate functions of E mapping to $\Pi_\alpha R_\alpha$ is the restriction to $\overset{\circ}{\Delta}$ of a map $E^\alpha(x)(k)$ with α some locus and k some gene at the α locus. By (0.1) these functions are all of the form $E^a(x)$ in Table 4.

The different functions L^S_{ij} are not independent of one another. For example, the simplest relations among them are:

(5.4)
$$L^S_{ij} = L^{\tilde{S}}_{ij} = -L^S_{\overline{ij}}.$$

There are more subtle linear relations as well. By a direct argument, Shahshahani showed in [28] that there are $d = n - \Sigma_\alpha n_\alpha + \ell - 1$ independent functions L^S_{ij}. Putting them together we get a function $L: \overset{\circ}{\Delta} \to R^d$. This number is exactly the number of dimensions left over after we constrain by the map E. This motivates the following theorem of Chapter II.

<u>3 Theorem</u>: The map $E \times L: \overset{\circ}{\Delta} \to (\Pi_\alpha \overset{\circ}{\Delta}_\alpha) \times R^d$ is a diffeomorphism, that is, it is one-to-one and onto with a smooth inverse map. Furthermore if E^a is any of the $\Sigma\, n_\alpha$ coordinate functions of E and L^b is any function L^S_{ij} (for example, the coordinate functions of L) then the gradients $\overline{\nabla} E^a$ and $\overline{\nabla} L^b$ are everywhere orthogonal with respect to the Shahshahani metric.

<u>Sketch of the Proof</u>: The orthogonality relations are consequences of (4.13). From them it is not heard to show that the derivative of $E \times L$ is a linear isomorphism at each point. The inverse function theorem (Thm. 4.3) then implies that $E \times L$ is locally a diffeomorphism. This means that if the inverse exists it is smooth. A final topological argument shows that the function is globally one-to-one and onto. QED

A <u>foliation</u> of a manifold is a way of cutting the manifold up into infinite disjoint family of lower dimensional submanifolds, called the <u>leaves</u> of the foliation. For example, by choosing a plane in R^3

we can foliate R^3 by the family of all planes parallel to the given one. By bending this picture you see a general picture of what a two dimensional foliation of a three dimensional manifold looks like (at least locally). From the functions E and L we get two foliations of $\overset{\circ}{\Delta}$ originally defined by Shahshahani.

The foliation of fibres $\overline{\mathfrak{D}}$ consists of the manifolds defined implicitly by a choice of marginal distributions, i.e. a typical leaf is of the form $E^{-1}(\{p^\alpha\})$ for $\{p^\alpha\}$ a fixed element of $\Pi_\alpha \overset{\circ}{\Delta}{}^\alpha$. Since E is linear the leaves of $\overline{\mathfrak{D}}$ are parallel convex sets in $\overset{\circ}{\Delta}$.

The transverse foliation $\overline{\mathcal{J}}$ consists of the manifolds defined implicitly by a choice of values for L. A typical leaf is of the form $L^{-1}(u)$ where u is a fixed vector in R^d. Since the functions L^S_{ij} are nonlinear these leaves are curved.

Every point p of $\overset{\circ}{\Delta}$ is the intersection of a unique leaf $\overline{\mathfrak{D}}_p$ of $\overline{\mathfrak{D}}$ (namely, $E^{-1}(E(p))$) and a unique leaf $\overline{\mathcal{J}}_p$ of $\overline{\mathcal{J}}$ (namely, $L^{-1}(L(p))$). $\overline{\mathfrak{D}}_p$ consists of all (interior) distributions having the same marginals as p. $\overline{\mathcal{J}}_p$ consists of all distributions having the same linkage numbers L^S_{ij} as p. In particular, Cor. 2 implies that the Wright manifold in $\overset{\circ}{\Delta}$, $\overset{\circ}{\Lambda}$, is a leaf of $\overline{\mathcal{J}}$. In fact, $\overset{\circ}{\Lambda} = L^{-1}(0)$.

While we have defined the leaves implicitly, each one can be described explicitly. On each leaf $\overline{\mathcal{J}}_p$ of $\overline{\mathcal{J}}$, L is a constant and so E restricts to a diffeomorphism of $\overline{\mathcal{J}}_p$ with $\Pi_\alpha \overset{\circ}{\Delta}_\alpha$. So the inverse function $(E|\overline{\mathcal{J}}_p)^{-1}$ if you could compute it gives an explicit coordinatization of $\overline{\mathcal{J}}_p$ by $\Pi_\alpha \overset{\circ}{\Delta}_\alpha$. For $\overset{\circ}{\Lambda}$ this function is given by the formula (2.8). Similarly, $(L|\overline{\mathfrak{D}}_p)^{-1}$ coordinatizes the leaf $\overline{\mathfrak{D}}_p$ by R^d. However, I am unable to actually compute these inverse functions.

The tangent space of a foliation at a point means the tangent space of the leaf through the point. So we write $T_p\bar{\mathcal{D}}$ or $T_p\bar{\mathcal{J}}$ and mean $T_p(\bar{\mathcal{D}}_p)$ or $T_p(\bar{\mathcal{J}}_p)$. We now state the result which relates this geometry to epistasis.

<u>4 Theorem</u>: (a) For $p \in \overset{\circ}{\Delta}$ the tangent spaces $T_p\bar{\mathcal{D}}$ and $T_p\bar{\mathcal{J}}$ give a perpendicular decomposition of $T_p\overset{\circ}{\Delta}$ with respect to the Shahshahani metric $(\ ,\ )_p$, i.e. the two subspaces are orthogonal and every vector X in $T_p\overset{\circ}{\Delta}$ can be written uniquely as the sum $X = X_d + X_t$ with $X_d \in T_p\bar{\mathcal{D}}$ and $X_t \in T_p\bar{\mathcal{J}}$.

(b) $T_p\bar{\mathcal{D}}$ is spanned by the vectors of the form $\bar{\triangledown}_p L^S_{ij}$.

(c) $X = \Sigma\ p_i\xi_i\partial_i \in T_p\overset{\circ}{\Delta}$ lies in $T_p\bar{\mathcal{J}}$ if and only if the relative component vector ξ has no epistasis, i.e. ξ can be written in the form (2.5).

This theorem interprets the first least squares approximations of Sec 2 geometrically. First, since X_t is perpendicular to X_d, Thm. 3.2 says that X_t is the projection of X on $T_p\bar{\mathcal{J}}$ orthogonal with respect to $(\ ,\ )_p$. Now if $X = \Sigma\ p_i\eta_i\partial_i$ and $X_t = \Sigma\ p_i\xi_i\partial_i$, Prop. 4.3 and (c) above imply that ξ is the zero-epistasis approximation to η with respect to the covariance inner product $_p(\ ,\ )$. The full power of these results is not so much in their interpretation of the static picture with p fixed as in their application to the dynamics of changing p.

<u>5 Corollary</u>: Let $X(p): \Sigma\ X_i(p)\partial_i = \Sigma\ p_i\xi_i(p)\partial_i$ be a vectorfield on $\overset{\circ}{\Delta}$ with $X_i(p)$ and $\xi_i(p)$'s smooth functions of p. The following conditions are equivalent:

($\bar{\mathcal{J}}$a) X is everywhere tangent to $\bar{\mathcal{J}}$, i.e. $X(p) \in T_p\bar{\mathcal{J}}$.

($\bar{\mathcal{J}}$b) The leaves of $\bar{\mathcal{J}}$ are invariant manifolds for the differential equation defined by X.

($\bar{\mathcal{J}}$c) For all i,j and S the functions L_{ij}^{S} are conserved (remain constant) by the differential equation defined by X.

($\bar{\mathcal{J}}$d) There exist $\Sigma\, n_{\alpha}$ smooth function $\varphi_{i_{\alpha}}^{\alpha}: \overset{\circ}{\Delta} \to R$ $(i_{\alpha} \in I_{\alpha})$ such that the $\Pi\, n_{\alpha}$ functions $\xi_i: \overset{\circ}{\Delta} \to R$ $(i \in I = \Pi_{\alpha}I_{\alpha})$ can be written $\xi_i = \Sigma_{\alpha}\varphi_{i_{\alpha}}^{\alpha}$.

The following conditions are equivalent:

($\bar{\mathfrak{D}}$a) X is everywhere tangent to $\bar{\mathfrak{D}}$, i.e. $X(p) \in T_p\bar{\mathfrak{D}}$.

($\bar{\mathfrak{D}}$b) The leaves of $\bar{\mathfrak{D}}$ are invariant manifolds for the differential equation defined by X.

($\bar{\mathfrak{D}}$c) The marginal distributions $p_{i_{\alpha}}^{\alpha}$ are conserved by the differential equation defined by X.

Since <u>frequency-dependent</u> selection is just some vectorfield X on $\overset{\circ}{\Delta}$ the condition ($\bar{\mathcal{J}}$a) - ($\bar{\mathcal{J}}$d) explain what zero-epistasis means for frequency-dependent selection. Since the L_{ij}^{S}'s are conserved their antilogs $p_ip_j/p_{\bar{i}}p_{\bar{j}}$, denoted z_{ij}^{S}, are also conserved. However, their relatives $d_{ij}^{S} = p_ip_j - p_{\bar{i}}p_{\bar{j}}$ are not conserved except on $\overset{\circ}{\Lambda}$ where all of the d_{ij}^{S}'s and all of the L_{ij}^{S}'s vanish.

Now apply Prop. 3.3:

6 Corollary: Let f: $\overset{\circ}{\Delta} \to R$ be a smooth function. The gradient field $\bar{\triangledown}f$ is everywhere tangent to $\bar{\mathcal{J}}$, i.e. $\bar{\triangledown}_pf \in T_p\bar{\mathcal{J}}$ for all $p \in \overset{\circ}{\Delta}$, if and only if f is constant on the leaves of $\bar{\mathfrak{D}}$. These conditions hold if and only if f depends only on the marginal distributions, i.e. there exists a smooth function $f_0: \Pi_{\alpha}\overset{\circ}{\Delta}_{\alpha} \to R$ such that f is the composition $f_0 \cdot E$.

Similarly, $\bar{\triangledown}f$ is everywhere tangent to $\bar{\mathfrak{D}}$ if and only if f depends only on the L_{ij}^S's.

These results all generalize to higher levels of epistasis. Let K be a fixed complex of loci as defined in Sec. 2. We define E^K to be the restriction to Δ of the product of the maps E^S for $S \in K$ where E^S is defined by (0.2). So when we know $E^S(p)$ we not only know the probabilities of the various genes but also the linkage among loci in S where S is any bloc in K. E^S maps Δ linearly onto some convex set denoted Δ_K and maps $\overset{\circ}{\Delta}$ onto its interior $\overset{\circ}{\Delta}_K$. Again the component functions are of the form $E^a(p)$ and there exist functions of the form $L^b(p)$ which play the role of the L_{ij}^S's. These functions are computable but they get pretty messy. Choosing a maximal independent set we define a function $L^K: \overset{\circ}{\Delta} \to R^{d(K)}$. Again $E^K \times L^K: \overset{\circ}{\Delta} \to \overset{\circ}{\Delta}_K \times R^{d(K)}$ is a diffeomorphism. The gradients of the E and L coordinate functions are orthogonal. So we get two orthogonal foliations $\bar{\mathfrak{D}}^K$ and $\bar{\mathfrak{J}}^K$. Most important, $X = \Sigma \ p_i \xi_i \partial_i \in T_p \overset{\circ}{\Delta}$ lies in $T_p \bar{\mathfrak{J}}^K$ if and only if the relative component vector ξ has K type epistasis. Apply these results with K equal to increasing skeleta of L: $L^{(1)}$, $L^{(2)}$, etc. We get the geometrical interpretation of the partition of variance in Sec 2. As K increases the leaves of $\bar{\mathfrak{J}}^K$ get thicker (increase in dimension) and the leaves of $\bar{\mathfrak{D}}^K$ get thinner to compensate.

The analogue of Cor. 5 also holds as does:

7 Corollary: Let f: $\overset{\circ}{\Delta} \to R$ be a smooth function. The following are equivalent:

(a) The gradient field $\bar{\nabla}f$ is everywhere tangent to $\bar{\mathfrak{J}}^K$.

(b) The components of the usual gradient: $\dfrac{\partial f}{\partial x_i}$ have K type epistasis at every point p of $\overset{\circ}{\Delta}$.

(c) f is constant on the leaves of $\bar{\mathfrak{D}}^K$.

(d) f depends only on the partial distributions $p^S \in \Delta_S$ for

$S \in K$, i.e. there exists $f_K \colon \overset{\circ}{\Delta}_K \to R$ such that $f = f_K \cdot E^K$.

6. The Selection Field

Mean fitness is the function $\bar{m} \colon \Delta \to R$ defined by $\bar{m} = \Sigma \ p_i p_j m_{ij}$

where m_{ij} is the fitness or Malthusian parameter of zygotic genotype

ij. $\bar{m}$ extends to the quadratic function $\Sigma \ x_i x_j m_{ij}$ on R^I. Using

this extension a direct computation using (4.12) shows that the $\overset{\circ}{\Delta}$

gradient of $\frac{1}{2} \bar{m}$ with respect to the Shahshahani metric is given by:

(6.1) $$\bar{\nabla}_p (\frac{1}{2} \bar{m}) = \sum p_i (m_i - \bar{m}) \partial_i \qquad p \in \overset{\circ}{\Delta}.$$

Here $m_i = \Sigma \ p_j m_{ij}$ is the average fitness of gametic genotype i.

The differential equation associated with this vectorfield on

$\overset{\circ}{\Delta}$ is (1.4). So we call $\bar{\nabla}(\frac{1}{2} \bar{m})$ the selection field on $\overset{\circ}{\Delta}$. The

observation that selection field is the gradient of mean fitness

(times 1/2) is Shahshahani's instant proof of the Kimura maximum

principal.

For Fisher's fundamental theorem let a_{ij} be any metric trait

depending on the zygotic genotype. $\bar{a} = \Sigma \ p_i p_j a_{ij}$ is the mean when

the gamete distribution is at $p \in \Delta$, and $a_i = \Sigma \ p_j a_{ij}$ is the mean or

average value of the trait when one gemete in the zygote is i. The

change in $\bar{a}$ as p changes due to the selection field is given by:

(6.2) $$\frac{d\bar{a}}{dt}\Big|_p = d_p \bar{a} (\bar{\nabla}_p (\frac{1}{2} \bar{m})) = (\nabla_p (\frac{1}{2} \bar{m}), \bar{\nabla}_p \bar{a})_p = 2 \sum p_i (m_i - \bar{m})(a_i - \bar{a})$$

$$= 2 \ \text{Cov}_p (m_i, a_i).$$

In detail, if p(t) is a solution curve of (1.4) then the derivative of $\bar{a}(p(t))$ with respect to t is equal to the derivative of the function $\bar{a}$ applied to the tangent vector p'(t). This is the chain rule. But by (3.25) the tangent vector p'(t) is just the vectorfield of the differential equation, in this case $\bar{\triangledown}(\frac{1}{2}\,\bar{m})$, at p(t). The remaining equations follow from the definition of gradient, (3.26), and the computation of $\bar{\triangledown}\bar{a}$ just like (6.1). In sum, the rate of change of the mean of a metric trait under selection alone is given by the genic covariance of the trait with fitness. When $a_{ij} = m_{ij}$ we have:

$$(6.3) \qquad \frac{d\bar{m}}{dt}\Big|_p = 2\left\|\bar{\triangledown}_p(\tfrac{1}{2}\,\bar{m})\right\|_p^2 = 2\sum p_i(m_i - \bar{m})^2 = 2\,\mathrm{Var}_p(m_i).$$

This is positive except where the vectorfield itself vanishes, i.e. except at equilibria.

The expression on the right of (6.3) is usually called the additive variance of fitness.

Suppose there is only one locus with n alleles. We say that fitness is <u>additive</u> or dominance-free if m_{ij} is the sum of a contribution from each gamete, i.e.

$$(6.4) \qquad\qquad\qquad m_{ij} = k_i + k_j.$$

Notice that in Sec. 2 we used additivity to mean the absence of interaction between loci in the expression of some gametic trait, i.e. no epistasis. Here we refer to the absence of interaction between homologous genes in the expression of a zygotic trait, i.e. no dominance. If (6.4) holds then letting $\bar{k} = \Sigma\, p_i k_i$, we have:

(6.5)
$$\bar{m} = 2\,\bar{k}$$

$$m_i = k_i + \bar{k}$$

$$m_i - \bar{m} = k_i - \bar{k}.$$

If m_{ij} does have dominance then with $p \in \Delta$ we can define the best dominance-free approximation to m_{ij} to be $m_i + m_j - \bar{m}$. Note that these values depend on p which we regard as fixed. So if we let $\theta_{ij} = m_{ij} - m_i - m_j + \bar{m}$, we have

(6.6) $\quad m_{ij} = \bar{m} + (m_i - \bar{m}) + (m_j - \bar{m}) + \theta_{ij} \quad$ (p fixed in Δ).

This fits into the linear algebra framework of Thm. 3.2. $\{p_i p_j\}$ defines a product distribution on the set of (ordered) zygotic zenotypes $I \times I$. $m_i + m_j - \bar{m}$ is the projection of the vector m_{ij} on the set of dominance-free vectors. The projection is orthogonal with respect to the inner product analogous to $_p(\ ,\ )$ i.e. $(m_{ij}, n_{ij}) = \Sigma\, p_i p_j m_{ij} n_{ij}$. This orthogonality means the following formula which holds because $\theta_i = \Sigma\, p_j \theta_{ij} = 0$ for all i:

(6.7) $$p_i p_j (a_i + b_j) \theta_{ij} = 0\,(a,b \in R^I).$$

As a consequence of this orthogonality, if we define the zygotic variance of fitness:

(6.8) $$V_Z = \sum p_i p_j (m_{ij} - \bar{m})^2$$

Then V_Z is the sum of the variance of the dominance-free approximation and the variance of the error term θ_{ij}. The first is called the additive variance and the second is called the dominance variance. Since $\bar{\theta} = 0$, we have:

$$V_Z = V_A + V_D$$

(6.9)
$$V_A = 2 \sum p_i (m_i - \bar{m})^2$$

$$V_D = \sum p_i p_j \theta_{ij}^2.$$

Thus, the additive part of the zygotic variance of fitness is twice the gametic variance.

The simplicity of these formulae comes from the fact that the zygotic distribution is a product. So the Hardy-Weinberg condition here plays the role that linkage equilibrium plays in problems of epistasis. When the Hardy-Weinberg condition does not hold the formulae for the dominance-free approximation becomes more complicated (see [6, Sec. 4.1]).

Now suppose that m_{ij} was dominance-free to begin with, i.e. (6.4) holds. Then $\frac{1}{2} \bar{m} = \bar{k}$ is a linear function. If all of the k_i's are distinct then $\bar{k}$ has a unique maximum on Δ namely fixation at the allele with the largest k_i, and so the selection field $\bar{\nabla k}$ tends to that vertex in Δ. In this case, however, we can explicitly solve system (1.4):

(6.10)
$$\frac{d \, \ell n(p_i/p_j)}{dt} = k_i - k_j \qquad i,j \in I.$$

Since $k_i - k_j$ is a constant, we get that p_i/p_j increases exponentially at rate $k_i - k_j$. So the vector $p(t) = \{p_i(t)\}$ is proportional to $\{p_i(0) e^{k_i t}\}$ or:

(6.11)
$$p_i(t) = p_i(0) e^{k_i t} / [\Sigma \, p_j(0) e^{k_j t}].$$

This case illustrates a difficulty with the Shahshahani metric. It is only defined for the interior distributions $\overset{\bullet}{\Delta}$. Often however the boundary $\Delta - \overset{\bullet}{\Delta}$ is exactly where we are looking, in problems about fixation of one allele, for example. Theorem 4.1 provides a way around this limitation:

1 Theorem: Let $f: S_2 \to \Delta$ by $f(z) = p$ with $p_i = z_i^2/2$ be the map of Thm. 4.1. Define $M: S_2 \to R$ by $M(z) = \Sigma \, z_i^2 z_j^2 m_{ij}$. With the usual metric on S_2 the gradient vectorfield $\triangledown(M/8)$ defines a differential equation on S_2 and f maps the solutions of this differential equation to solutions of the selection equation (1.4) on Δ.

Proof: On $S_2 \cap \overset{\bullet}{P}$ f is an isometry and so its derivative relates the gradient of $M/8$ with $\overline{\triangledown}(\frac{1}{2} \, \overline{m})$ on $\overset{\bullet}{\Delta}$. By continuity the derivative relates $\triangledown(M/8)$ on the rest of $S_2 \cap P$ with equation (1.4) on the rest of Δ. The extension to all of S_2 has to do with the invariance of M under change of sign of the coordinates z_i. This is a technical argument which we will sketch below. If two vectorfields are related by the derivative of f they are called f related and it then follows from the chain rule that f maps solutions of one to solutions of the other.   QED

This result is a special case of a general method of changing problems at the boundary of Δ to problems with symmetry on S_2. Here I have to become technical. The group $(Z_2)^n$ acts on S_2 by $e_i(z_1,\ldots,z_n) = (z_1,\ldots,-z_i,\ldots,z_n)$. The quotient of S_2 under this action is Δ and $f: S_2 \to \Delta$ is essentially the quotient map. The general form of a vectorfield on Δ which is parallel to the faces is $X(p) = \Sigma \, p_i \xi_i(p)\partial_i$ with $p_i \overline{\xi}_i(p) = 0$. This is the general form of

frequency dependent selection and the parallelism condition means
that new types of gametes cannot be produced by selection above. In
contrast, recombination and mutation do produce new types of gametes
and don't satisfy the parallel condition. Such a vectorfield is f
related to a unique vectorfield $Y(z) = \Sigma z_i \eta_i(z) \partial_i$ on S_2 which Y is
invariant under the group action. Conversely such a vectorfield Y
folds up to get a unique vectorfield X on Δ parallel to the faces.
Finally, since the functions η_i are invariant under the group action
it is an exercise in singularity theory [3, p. 64, exercise 1] to show
that η_i depends on the squares of the z_i's. So the study of frequency
dependent selection on Δ is equivalent to the study of vectorfields
on S_2 invariant under the action of $(Z_2)^n$. However, we won't pursue
these boundary problems further here.

Since the selection field is a gradient Cors. 5.6 and 5.7
apply to it. The result is a bit sharper in the completely symmetric
case. Recall from Sec. 2 that m_{ij} is completely symmetric if
$m_{ij} = m_{\overline{ij}}$ for all i,j ϵ I and $S \subset L$.

2 Theorem. Suppose m_{ij} is completely symmetric. The following equivalent conditions define the absence of epistasis for the selection
field $\overline{\triangledown}(\frac{1}{2} \overline{m})$:

(a) The selection field is everywhere tangent to the transverse foliation $\overline{\mathcal{J}}$.

(b) Each leaf of $\overline{\mathcal{J}}$ is an invariant manifold for solutions
of the differential equation (1.4).

(c) For all i,j ϵ I and $S \subset L$, L_{ij}^S is conserved under
select on.

(d) Mean fitness at p depends only on the marginal (i.e.

gene) distributions p^α.

(e) Fitness is the sum of contributions from the separate

loci, i.e. for $\alpha = 1,\ldots,\ell$ there exist symmetric, real vectors

$m^\alpha \in R^{I_\alpha \times I_\alpha}$ such that

(6.12)
$$m_{ij} = \sum_\alpha m^\alpha(i_\alpha, j_\alpha).$$

3 Corollary: If the selection field has no epistasis then the Wright

manifold of distributions in linkage equilibrium is an invariant mani-

fold for selection.

The above result generalizes to define K-type epistasis by:

(a) $\bar{\nabla}(\frac{1}{2}\bar{m})$ leaves $\bar{\mathcal{J}}_K$ invariant, (b) $\bar{m}$ at p depends on the par-

tial distributions p^S for S in K i.e. mean fitness depends only on

the joint distribution on blocs of loci in K, and (c) m_{ij} can be

written as the sum of terms $m^S(i_S, j_S)$ for S in K, terms each

depending only on the genes in a bloc of K.

For any K there is a particular leaf Λ_K of $\bar{\mathcal{J}}_K$ which is anal-

ogous to the Wright manifold $\overset{\circ}{\Lambda}$ in $\bar{\mathcal{J}}$. For any leaf of the fibre

foliation $\bar{\mathfrak{D}}_K$ the point of intersection with Λ_K is the distribution of

greatest randomness in the $\bar{\mathfrak{D}}_K$ leaf. If every locus of L is in some

bloc of K then Λ_K contains $\overset{\circ}{\Lambda}$. While in the presence of epistasis

$\overset{\circ}{\Lambda}$ is not invariant, if the epistasis is K type then Λ_K is invariant.

The most important special case of K is the disjoint bloc

model $K = T_1 \vee \ldots \vee T_{\ell'}$ of Sec. 2. Here Λ_K consists of the distri-

butions satisfying (5.3) for $S = T_1, T_2, \ldots, T_{\ell'}$. By an argument

analogous to Cor. 5.2, $p \in \Lambda_K$ for $K = T_1 \vee \ldots \vee T_{\ell'}$ if and only if p

is the product distributions from the factors $I_{T_1}, \ldots, I_{T_{\ell'}}$:

(6.13)
$$P_i = p^{T_1}(i_{T_1}) \ldots p^{T_{\ell'}}(i_{T_{\ell'}}).$$

7. The Recombination Field.

Following equation (1.6) we define the recombination field associated to $S \subset L$ by:

(7.1)
$$R^S = \sum_{i,j} (r_{ij}^S b_{ij} P_i P_j - r_{\bar{i}\bar{j}}^S b_{\bar{i}\bar{j}} P_{\bar{i}} P_{\bar{j}}) \partial_i.$$

Recall that $\bar{i}$ and $\bar{j}$ are $i_S j_{\tilde{S}}$ and $j_S i_{\tilde{S}}$ respectively. The differential equation (1.6) comes from the vectorfield $-R$ where

(7.2)
$$R = \sum \{R^S : S \subset L\}.$$

Shahshahani observed that the recombination fields are all "vertical" vectorfields, that is, they are everywhere tangent to the fibre foliation $\tilde{\mathfrak{D}}$ of Sec. 5. To see this, compute the gradient of L_{ij}^S defined by (5.2). Using Table 4 of Sec. 4 we get:

(7.3)
$$\bar{\nabla} L_{ij}^S = \partial_i - \partial_{\bar{i}} - \partial_{\bar{j}} + \partial_j.$$

Now fix i,j and S and collect the eight terms in the formula for R^S which involve the pairs ij and $\bar{i}\bar{j}$. We discover that $r_{ij}^S b_{ij} P_i P_j$ occurs as the coefficient of ∂_i, ∂_j, $-\partial_{\bar{i}}$ and $-\partial_{\bar{j}}$ while $r_{\bar{i}\bar{j}}^S b_{\bar{i}\bar{j}} P_{\bar{i}} P_{\bar{j}}$ occurs as the coefficient of $-\partial_i$, $-\partial_j$, $\partial_{\bar{i}}$ and $\partial_{\bar{j}}$. So we get:

(7.4)
$$R^S = (1/4) \sum_{i,j} (r_{ij}^S b_{ij} P_i P_j - r_{\bar{i}\bar{j}}^S b_{\bar{i}\bar{j}} P_{\bar{i}} P_{\bar{j}}) \bar{\nabla} L_{ij}^S.$$

The factor of 1/4 is to compensate for the four identical terms here associated with the pairs ij, ji, $\overline{i}\overline{j}$ and $\overline{j}\overline{i}$.

Since the recombination fields R^S and R are everywhere linear combinations of $\overline{\nabla}L_{ij}^S$, Thm. 5.4 (b) implies that these fields are everywhere tangent to $\overline{\mathfrak{D}}$, i.e. satisfy $(\overline{\mathfrak{D}}a)$ - $(\overline{\mathfrak{D}}c)$ of Cor. 5.5. This just says that the gene frequencies at each locus are left unchanged by recombination.

As we saw in equation (1.7) the form of the recombination fields is simplified when r_{ij}^S and b_{ij} are completely symmetric. In that case we have (cf. (5.1)):

(7.5)
$$R^S = \sum_{i,j} r_{ij}^S b_{ij} (p_i p_j - p_{\overline{i}} p_{\overline{j}}) \partial_i$$

$$= (1/4) \sum_{i,j} r_{ij}^S b_{ij} (p_i p_j - p_{\overline{i}} p_{\overline{j}}) \overline{\nabla}L_{ij}^S.$$

$$= (1/4) \sum_{i,j} r_{ij}^S b_{ij} d_{ij}^S \; \overline{\nabla}L_{ij}^S.$$

As Prop. 5.1 indicates d_{ij}^S and L_{ij}^S are measuring essentially the same thing. Notice that if we replace d_{ij}^S by L_{ij}^S in (7.5) we get constant linear combinations of terms like:

(7.6)
$$L_{ij}^S \overline{\nabla}L_{ij}^S = \overline{\nabla}(\frac{1}{2}(L_{ij}^S)^2).$$

So the resulting vectorfield is a gradient field. However, d_{ij}^S is not a function of the L_{ij}^S's above and so R^S is not a gradient field even in the completely symmetric case. We will see in Sec. 9 that this has profound consequences for the dynamics of selection plus recombination.

Despite the fact that the fields R^S are not gradients, there does exist a global Lyapunov function for the recombination field -R on $\overset{\bullet}{\Delta}$. It is entropy H:

(7.7)
$$H(p) = -\sum p_i \, \ell n \, p_i.$$

More precisely, we normalize H by subtracting the sum of the entropies of the gene frequencies:

(7.8)
$$\hat{H}(p) = H(p) - \sum_{\alpha} H(p^{\alpha}).$$

Recall that p^{α} is the marginal distribution induced by p on the alleles I_{α} at the α locus.

$\hat{H}$ turns out to be everywhere nonpositive on $\overset{\bullet}{\Delta}$, vanishing precisely on the Wright manifold $\overset{\bullet}{\Lambda}$.

<u>1 Theorem</u>: Assume r_{ij}^{S} and b_{ij} are completely symmetric. Let p(t) be a path in $\overset{\bullet}{\Delta}$ associated to the recombination vectorfield - R. That is, p(t) is a solution of the recombination equation (1.7).

(7.9)
$$\frac{d\hat{H}}{dt}\Big|_p = (-R, \bar{\triangledown}_p \hat{H})_p \geq 0 \qquad \text{for all} \quad p \in \overset{\bullet}{\Delta}.$$

Furthermore, $(-R, \bar{\triangledown}_p \hat{H})_p$ vanishes at p if and only if the vectorfield R is zero at p, i.e. p is an equilibrium for equation (1.7). R vanishes on the Wright manifold and $\bar{\triangledown}\hat{H}$ vanishes exactly on the Wright manifold.

We prove this theorem in Chap. III together with a related result for certain kinds of position effects.

One point left open in the theorem is the question whether R can vanish off the Wright manifold, i.e. are there any equilibria other than distributions satisfying (2.8). Assuming that $r_{ij}^S b_{ij} > 0$

for all i,j and S R vanishes precisely on Λ. However, if too

many b_{ij}'s are 0, aberrant equilibria can occur. I suspect that they

are not biologically significant. If too many zygote types are

sterile I conjecture that the population either becomes extinct or

else eliminates certain gamete types thus throwing the dynamics onto

some lower dimensional "face" of Δ on which only the usual linkage

equilibria occur. This <u>Sterility</u> <u>Conjecture</u> is also described in

Chap. III.

In relating recombination with selection, I originally hoped

that for each complex of loci K, the recombination field would be

tangent to the distinguished leaf Λ_K of the transverse foliation $\bar{\mathcal{J}}_K$.

This would then imply that if the selection field had K-type epistasis,

Λ_K would be an invariant manifold for selection plus recombination.

However, this tangency condition is usually not true even when K is

a disjoint bloc model $T_1 \vee \ldots \vee T_\ell$. At least in the disjoint bloc

case this is a shocking result.

In that case Λ_K consists of distributions $p \in \overset{\circ}{\Delta}$ satisfying

(6.13) meaning that the loci on different blocs may be independent.

There is an example where it represents a considerable violation of

biological intuition. We will examine this case in some detail since

at first glance it casts doubt on the validity of the entire model.

Consider the case when there are ℓ' chromosomes in the haploid

gamete and the bloc T_a consists of the loci on the ath chromosome

$(a = 1, \ldots, \ell')$. The distributions in Λ_K are precisely those where

the loci on different chromosomes are independently distributed. Now

it is not surprising that selection might lead to "linkage" between genes on different chromosomes, but we will see in Chap. III that even if the birth rates, death rates and recombination rates are genotype independent, i.e. there is no selection, R need not be tangent to Λ_K. So recombination alone may destroy the independence between chromosomes. But biological intuition going back to Mendel's Law of Independent Assortment makes it hard to see how such independence could be destroyed at all and especially by recombination which tends to make all the loci independent. Perhaps the model is wrong and a more accurate model would yield tangency in the disjoint bloc case. But no, the model is right.

The explanation is simpler for a discrete time model. Of these there are two types: In the first kind, the annuals, this year's crop all die but are replaced by their offspring. In the second kind, the perennials, some of this year's crop die and the rest survive to join the offspring in forming next year's crop. If the model ignores the age structure of the population then the survivors and the young are regarded as indistinguishable members of the new crop. In that case the response to selection alone is essentially the same in the two kinds of model. All that matters are the net reproductive rates (= birth rates minus death rates). In the biological literature there is a large body of work about life-history tactics much of which has grown out of Cole's observation [5] that the switch from an annual strategy to a perennial strategy (in fact to immortality) is mathematically equivalent to increasing the birth rate by one. But once one introduces recombination the two models are quite different, for the offspring have been exposed to the effects of recombination while the survivors have not. The new crop is thus a mixture of two

populations. It is an odd fact that two characters can be mathematically independent in each of two populations and yet not be independent in the combined population.

For a concrete example, consider a four locus, two allele model
with two loci on each chromosome. Imagine a large population with
very small per generation birth and death rates (and these are independent of the genotypes). Suppose that the population begins with
A, B and a, b totally linked and in equal numbers on the first
chromosome and with C, D and c, d totally linked in equal numbers on
the second chromosome. Finally, assume that the two chromosomes are
independently distributed. Thus, out of the 16 possible gamete-types
the parental generation was constructed out of the four possibilities.
AB or ab on the first chromosome and CD or cd on the second. The
probability of each of the four types is 1/4. Thus, the parental
generation is far from linkage equilibrium, Λ, but is on Λ_K for the
disjoint bloc model with the two blocs equal to the two chromosomes.
Now look one year later supposing recombination rates r_1 and r_2 within the two chromosomes. Among the offspring all 16 gamete-types
appear and the offspring distribution is closer to Λ (though not on
Λ unless $r_1 = r_2 = 1/2$). Furthermore, since the chromosomes assort
independently the offspring distribution is on Λ_K. Meanwhile, the
distribution of the survivors is the same as the parental distribution, since the death-rate is genotype independent, and so it is
still on Λ_K. However, for the entire population of survivors-plus-
offspring the two chromosomes are not independent. For choose a gamete and suppose we discover Ab on the first chromosome. Then the
gamete must come from the offspring population and the probability

that Cd is on the second chromsome is $r_2/4$. On the other hand if we discover AB on the first chromosome then since the number of survivors is very large and the number of offspring is very small, it is almost certain that the gamete comes from the survivor population and so the probability that Cd is on the second chromosome is essentially 0. Since the state of the first chromosome provides information about the state of the second, the two chromosomes are not independent.

The vectorfield model is like a discrete model of the second kind and the constant mixing of survivor and offspring distributions is the reason that tangency fails.

8. The Mutation Field.

From equation (1.8) we define the mutation field:

(8.1)
$$N = \sum p_i N_{ij} \partial_j .$$

Here N_{ij} is defined to be the matrix:

(8.2)
$$N_{ij} = \begin{cases} n_{ij} & i \neq j \\ -n_{i*} \equiv -\sum_k n_{ik} & i = j. \end{cases}$$

n_{ij} is the relative rate at which i gametes are transformed to j gametes by mutation ($i \neq j$). Unlike the recombination and selection fields, N is a linear vectorfield. The coefficient of ∂_j is a linear function of p. Furthermore, $N_{ij} = n_{ij} \geq 0$ when $i \neq j$. We make the reasonable assumption that any gamete j can be obtained from any other gamete i by some finite sequence of mutational steps. Under this assumption, the theory of positive matrices due to Frobenius enables us to prove--in Chap. III--that the vectorfield N vanishes at a unique point q in Δ. q lies in $\overset{\bullet}{\Delta}$ and can be directly computed. Up to scalar multiple--so that $\Sigma \, q_i = 1$ -- q_i is the ith principal minor of N (= the determinant of the n-1 × n-1 matrix obtained by deleting the ith row and the ith column from N). This equilibrium is globally <u>asymptotically</u> <u>stable</u> meaning that every solution path of (1.8) approaches q as time t tends to ∞. Furthermore, we can estimate the eigenvalues of N to get an upper bound on the rate at which the equilibrium is approached, namely no faster than twice the total mutation rate $\max_i n_{i*}$. As the mutation rates are usually quite small the mutation equilibrium is approached in a rather leisurely fashion as one would expect.

Now suppose that mutation between gametes is due to the independent occurrence of mutations at each locus. At the α locus, let $n^{\alpha}_{i_\alpha j_\alpha}$ be the relative rate by which i_α alleles are transformed to j_α alleles by mutation when $i_\alpha \neq j_\alpha$. We assume that the rate at each locus is independent of the allelic values at the remaining loci. We then get a matrix $\bar{N}^{\alpha}$ and an associated linear vectorfield on Δ_α which we will also denote by $\bar{N}^{\alpha}$.

On Δ the vectorfield representing mutation at the α locus is the linear vectorfield N^{α} associated with the matrix defined by

(8.3) $$N^{\alpha}_{ij} = \bar{N}^{\alpha}_{i_\alpha j_\alpha} \, \delta_{i_{\tilde{\alpha}} j_{\tilde{\alpha}}}$$

Here δ is the Kronecker delta and $i_{\tilde{\alpha}}$ is the projection of i to the set of loci complementary to α, or to $\tilde{\alpha} = L - \{\alpha\}$. This means that i mutates to j at relative rate $n^{\alpha}_{i_\alpha j_\alpha}$ provided that i and j agree at all loci other than the α locus and at that locus they

are different. The vectorfield N representing the effect of mutation at all loci is the sum $N = \Sigma_\alpha N^\alpha$. It is the linear vectorfield
associated with the sum of the corresponding matrices.

When N has this special form the Wright manifold Λ is an
invariant manifold for N and the equilibrium q lies in Λ. In
fact, q is the product distribution whose marginal distribution q^α
is the equilibrium in $\overset{\circ}{\Delta}_\alpha$ for $\overline{N}^\alpha$. The rate of approach to q is at
most the minimum of the corresponding rates for the q^α's, expressing
the "bottleneck" effect whereby the rate of a complex process is
determined by the rate of the slowest component.

Notice that in defining N we did not include any terms representing simultaneous mutation at several loci. Why don't such terms
appear?

The reason goes back to the definition of mutation rate.
Beginning with $i \in I$, the probability that in a time interval of length
dt i will mutate to j which agrees with i at all loci but α
is $p_i n^\alpha_{i_\alpha j_\alpha} dt$, or more precisely it is this plus an error term of the
form $o(dt)$ meaning $\text{Lim } o(dt)/dt = 0$ as dt approaches 0. Similarly
the probability that i will mutate to k which agrees with i at
all loci but $\beta \neq \alpha$ is $p_i n^\beta_{i_\beta k_\beta} dt + o(dt)$. Now if ℓ agrees with j
at α, with k at β and i everywhere else then the probability
the i will mutate to ℓ directly in the dt interval is the product,
$p_i^2 n^\alpha_{i_\alpha j_\alpha} n^\beta_{i_\beta k_\beta} dt^2 + o(dt^2)$, assuming that mutations at different loci
are independent. This entire term is $o(dt)$ and so the instantaneous
rate of such simultaneous mutations is zero.

9. The Combined Field.

By the combined field we mean the sum of the effects of selection, recombination and mutation:

$$(9.1) \qquad \bar{\nabla}(\tfrac{1}{2}\,\bar{m}) - R + N.$$

Ideally, we would like to solve explicitly the differential equation associated to this field. As this is impossible we would at least like to give a portrait of how the solutions behave, a "qualitative" description. We are particularly interested in the stable elements of behavior. There are two--quite different-- concepts of stability for a vectorfield on a manifold.

Stability of a particular solution under perturbation of initial conditions is called orbital stability. For example, an equilibrium is called stable if for initial conditions near it the solutions tend to return to the equilibrium or at least remain close to it. The solution paths which are not orbitally stable are called separatrices. Removing the separatrices cuts the manifold up into a number of invariant open sets or domains of attraction. All of the solutions in a domain behave similarly. For example, they may all tend to a particular equilibrium or limit cycle. Crossing a separatrix to a different domain changes the behavior, e.g. the solutions may now approach a different equilibrium. The location of the separatrices and description of the domains of attraction constitute the phase portrait of the field.

Stability of the entire equation under perturbation of parameters is called structural stability. If a vectorfield is not structurally stable then an arbitrarily small change in some coefficients or addition of some other small field (a perturbing term) may completely change the phase portrait of the equation. If the equation depends on some continuous parameter then the set of values of the parameter at which the equation is not structurally stable is called the bifurcation set. Removing the bifurcation set cuts the parameter space into a number of open sets or regimes. All of the equations with parameters in a regime have similar phase portraits. Crossing the bifurcation set to a different regime causes a change in phase portrait which Thom [34] calls a catastrophe.

The philosophical similarities between these two kinds of stability is discussed by Abraham in [1]. In practice some understanding of the qualitative behavior of an equation is needed even when doing numerical solutions. It is necessary to know that the finite set of points that one plots give a reasonable picture of the actual solution and that the solution itself is typical, i.e. is not an artifact due to special choice of initial values or coefficients.

Some of the difficulties that arise in studying the combined field come from the fact that recombination is never, and multi-locus selection is rarely, structurally stable when considered alone.

In Sec. 7, we saw that the recombination field -R is tangent to the foliation $\bar{\mathfrak{D}}$. A point moving along the recombination field remains in a fixed leaf, i.e. preserves the gene frequencies at each locus. It moves down the leaf toward linkage equilibrium approaching the unique point of the Wright manifold Λ in the original leaf. The tangency to a foliation and the occurrence of a whole submanifold of equilibria are structurally unstable characteristics. However, by invariant manifold theory (cf. [14]) the existence of the invariant

manifold itself is structurally stable. This means that if $-\tilde{R}$ is a
perturbation of $-R$, there exists a submanifold $\tilde{\Lambda}$ close to Λ which
is invariant with respect to $-\tilde{R}$. If a point begins far from Λ and
moves according to $-\tilde{R}$ it rapidly approaches the invariant manifold
$\tilde{\Lambda}$ with a slow change of gene frequencies at each locus. So in the
beginning it acts like a solution of $-R$. But near $\tilde{\Lambda}$ it begins to
move along $\tilde{\Lambda}$ in a fashion completely determined by the particular
perturbation, i.e. completely unpredictable from $-R$ alone. This slow
transverse motion which was invisible at first when $-R$ was large
becomes manifest near $\tilde{\Lambda}$ because there $-R$ itself is near zero.

For selection, suppose the maximum value of mean fitness $\bar{m}$
occurs at a point in $\overset{\circ}{\Delta}$. If this point is an isolated maximum then
the gradient $\bar{\nabla}(\frac{1}{2}\bar{m})$ is structurally stable. The point is an asymptotically stable equilibrium and a perturbation of the selection
field has a unique asymptotically stable equilibrium nearby. This
picture is reasonable in a single locus model. In a multilocus model
it corresponds to complete epistasis. At the other extreme suppose
that m_{ij} has zero epistasis. Then by Thm. 6.2 $\bar{m}$ is constant on
the leaves of $\bar{\mathfrak{D}}$. So the maximum value occurs on an entire leaf of
$\bar{\mathfrak{D}}$. Furthermore $\bar{\nabla}(\frac{1}{2}\bar{m})$ is everywhere tangent to the leaves of $\bar{\mathcal{J}}$ and
so a point moves along a fixed leaf of $\bar{\mathcal{J}}$ toward the maximum of $\bar{m}$
on the leaf. For K type epistasis, $\bar{\nabla}(\frac{1}{2}\bar{m})$ is tangent to the foliation $\bar{\mathcal{J}}_K$ and $\bar{m}$ is constant on leaves of $\bar{\mathfrak{D}}_K$. So here again we have
an invariant foliation and a submanifold of equilibria. Just as be-
fore this portrait is not structurally stable.

The mutation field N is structurally stable, but the coefficients are small enough that it is best considered just a perturba-

tion term added to $\bar{\nabla}(\frac{1}{2}\bar{m}) - R$.

In the zero-epistasis case the behavior of the combined field can be described completely. In that case the motion due to $-R$ of a point towards linkage equilibrium is never opposed by selection while the latter moves the point toward the maximum of mean fitness. The two motions are perpendicular. In particular,

$$\frac{d\bar{m}}{dt}\Big|_p = (\nabla_p(\frac{1}{2}\bar{m}) - R, \bar{\nabla}_p\bar{m})_p$$

(9.2)

$$= (\bar{\nabla}_p(\frac{1}{2}\bar{m}), \bar{\nabla}_p\bar{m})_p - 0 = V_A \geq 0.$$

So mean fitness increases monotonically. This is Ewens' extension [9] of Fisher's fundamental theorem to multilocus models when there is no epistasis.

The Wright manifold of linkage equilibria is an invariant manifold for the combined field. Using the diffeomorphism E: $\dot{\Lambda} \rightarrow \Pi_\alpha\dot{\Delta}_\alpha$ we can think of selection and mutation acting on Λ to affect the gene frequencies at each locus. Writing m_{ij} as the sum (6.12) and putting the analogue of the Shahshahani metric on each $\dot{\Delta}_\alpha$, the motion on Λ due to the combined field is separated via E into ℓ non-interacting fields: $\bar{\nabla}(\frac{1}{2}\bar{m}^\alpha) + \bar{N}^\alpha$ on Δ_α. If each $\bar{m}^\alpha$ has an isolated maximum in Δ_α and the mutation rates are small then these fields are structurally stable.

So in the zero epistasis case the combined field is usually structurally stable. Recombination tends to move the state to linkage equilibrium and selection plus mutation act to equilibrate the gene frequencies at each locus. Which of these two motions is faster depends upon the relative strengths of recombination and selection. The two cases give different patterns where epistasis is introduced.

Strong Recombination and Weak Selection: Here the combined field is a perturbation of -R. When a small epistasis component is included Λ is no longer invariant but there is an invariant manifold $\tilde{\Lambda}$ near Λ. Far from $\tilde{\Lambda}$ recombination is dominant and moves the point close to $\tilde{\Lambda}$ with slight change of gene frequencies. Near $\tilde{\Lambda}$ motion is determined by selection and mutation. E: $\tilde{\Lambda} \rightarrow \Pi_\alpha \Delta_\alpha$ is still a diffeomorphism but now the change in gene frequencies at the separate loci are not independent. There are small interaction terms due to epistasis and to the deviation from linkage equilibrium. This motion "parallel" to Λ is Shahshahani's description of quasi-linkage equilibrium [see [28, Sec. 5.4]).

Strong Selection and Weak Recombination: Suppose $\bar{m}$ has an interior maximum. If this maximum is isolated then the combined field acts like a perturbation of a one-locus, n-allele model. Recombination and mutation deflect the asymptotically stable equilibrium of the combined field away from the maximum of $\bar{m}$ unless this maximum happened to lie on Λ. If there is incomplete epistasis, say exactly K-type so that the maxima of $\bar{m}$ occur exactly on a leaf of $\bar{\mathfrak{D}}_K$, then selection alone would move a point toward this maximum leaf. The combined field has an invariant manifold $\tilde{D}$ near this maximum leaf which is bent toward Λ if the original did not meet Λ. Far from $\tilde{D}$ selection is dominant and moves the point close to $\tilde{D}$ with rapidly increasing mean fitness and only a slow change in the recombination values L_{ij}^S. Near $\tilde{D}$ motion is determined by recombination and mutation with little change in mean fitness. I conjecture that motion

on $\tilde{D}$ will move asymptotically toward an equilibrium near the point
of $\tilde{D}$ of maximum entropy.

We should point out where the "strength" of R comes from. It
is usually felt that this is a matter of linkage. Indeed, very tight
linkage, $r_{ij}^{S} \ll 1$, leads to weak recombination. But with moderate
linkage values the size of R tends to be determined by the birth-
rates, b_{ij}. In most natural populations most of the time mean fitness is about 0, i.e. exponential growth or decay must be transient.
So the fitness values m_{ij} tend to be clustered around 0. Now since
$m_{ij} = b_{ij} - d_{ij}$, a moderate fitness value m_{ij} can arise either through
moderate birth and death rates or through a high birth rate and a
compensating high death rate. The selection field doesn't distin-
guish between these two possibilities. But the latter case leads to
strong recombination relative to selection.

The pictures in these two extreme cases are quite different.
This suggests that the interesting intermediate cases might be quite
complicated. What can one say when selection and recombination are
of comparable strength? The best organizing principle is Wright's
"adaptive surface" point of view [36].

In a paper titled "On the Nonexistence of Adaptive Topographies"
[24] Moran constructed explicit examples in the two locus, two allele
case which showed that under the combined effect of selection and
recombination mean fitness need not be always increasing. With the
benefit of hindsight it is easy to construct a large class of such
examples.

Suppose that $\bar{m}$ has an isolated maximum at some point p of
$\mathring{\Delta} - \Lambda$. Now add in a small recombination term as in the second case

above. The combined field has at equilibrium $\tilde{p}$ near p but since $R \neq 0$ at p, $\tilde{p} \neq p$. p is an isolated maximum of $\bar{m}$, so $\bar{\triangledown}(\frac{1}{2}\bar{m})$ does not vanish at $\tilde{p}$. Different solution curves for the combined field will approach $\tilde{p}$ along different directions. If $\bar{m}$ is increasing for a particular solution, then along a solution approaching from the opposite direction $\bar{m}$ will be decreasing. If, for example, the matrix m_{ij} has distinct eigenvalues one can find enough approach directions to make this argument work.

But the title of Moran's paper makes too strong a claim. The adaptive surface point of view does not assume that evolution acts to maximize mean fitness, just that it tends to maximize some hypothetical "fitness function" [37]. For example, we have seen that recombination acts to increase entropy while selection acts to increase mean fitness. Perhaps the competing claims of selection and recombination can be measured by some combination of mean fitness and entropy which would increase along solutions of the combined field. The general hope leads to the following:

The Wright Conjecture: The combined field admits a global Lyapunov function. That is, there exists a "fitness function" $F: \Delta \rightarrow R$ such that $dF/dt > 0$ along solution paths other than equilibria.

Ewens raises essentially the same question [10, p. 96].

For zero epistasis fields the above program usually works. Note that $\bar{m}$ alone is not a Lyapunov function despite (9.2) because on the $\mathcal{D}$ leaf of maximum fitness the combined field moves the point toward Λ while $\bar{m}$ remains constant. However, addition of a small term related to entropy usually does give a Lyapunov function at least locally. The details are in Chap. IV.

Since the zero epistasis case is pretty clear without a
Lyapunov function, the real interest lies in the cases where there
is epistasis. In general, the Wright Conjecture is false because
periodic orbits or cycles can occur in such cases. Since a function
can't be continually increasing as one goes around a cycle the Wright
Conjecture fails in such cases. The existence of such cycles is
proved in Chap. IV by the demonstration that a Hopf bifurcation can
occur in the family of combined fields [22].

Certain kinds of bifurcation are consistent with the adaptive
surface picture and are familiar to geneticists. For example consider the family of real differential equations depending on the real
parameter λ:

(9.3) $$\frac{dx}{dt} = -x^3 + \lambda x = -x(x^2 - \lambda).$$

The vectorfield in (9.3) is the gradient of f_λ where

(9.4) $$f_\lambda(x) = -x^4/4 + \lambda x^2/2 = -x^2(x^2 - 2\lambda)/4.$$

When $\lambda \leq 0$, $x = 0$ is the only equilibrium of (9.3) and it is asymptotically stable. When $\lambda > 0$, $x = 0$ is an unstable equilibrium but
there are two asymptotically stable equilibria at $x = \pm\sqrt{\lambda}$. In terms
of the potential or fitness functions: For $\lambda \leq 0$ f_λ has only one peak,
or local maximum, and that is at 0. For $\lambda > 0$ there are two peaks
at $\pm\sqrt{\lambda}$ separated by a trough, or local minimum, at 0. Thus, as λ
changes from negative to positive passing the bifurcation point $\lambda = 0$,
we see the erosion and splitting of an adaptive peak into two. This
is the familiar metaphor for the early stage of speciation in the

adaptive surface picture.

But now consider the family of equations in R^2 depending on a real parameter λ:

(9.5)

$$\frac{dx}{dt} = \lambda x - y - x(x^2 + y^2)$$

$$\frac{dy}{dt} = x + \lambda y - y(x^2 + y^2).$$

In polar coordinates $r = x^2 + y^2$, $\tan \theta = y/x$ this family is:

(9.6)

$$\frac{d\theta}{dt} = 1$$

$$\frac{dr}{dt} = -r(r^2 - \lambda) \qquad (r \geq 0).$$

Since $d\theta/dt = 1$ every point (except the origin) moves counter-clock-wise about the origin with unit angular velocity. Now if $\lambda \leq 0$ dr/dt is always negative and so every point spirals inward to the origin. So the origin is an asymptotically stable equilibrium when $\lambda \leq 0$. But if $\lambda > 0$ dr/dt is positive for $r < \sqrt{\lambda}$ and so solutions beginning inside this circle spiral outward. Solutions outside the circle still spiral inward but not to an equilibrium. When $r = \sqrt{\lambda}$ $dr/dt = 0$ while $d\theta/dt = 1$ and so the motion along the circle of radius $\sqrt{\lambda}$ is a limit cycle solution of the equation. As in (9.3) the change of λ from negative to positive changes the stable equilibrium at the origin to an unstable equilibrium. But now instead of an adaptive peak splitting in two a limit cycle is emitted. This is the Hopf bifurcation.

The Hopf bifurcation does occur naturally in a biological con-text namely in predator-prey equations. In most forms of the predator-prey equations (one species each) there is a unique equilibrium point where both can coexist. Depending on the choice of parameters this equilibrium is either asymptotically stable or unstable. Here we are excluding the original Lotka-Volterra equation itself with its structurally unstable foliation of the first quadrant by cycles. When the equilibrium is unstable it is enclosed by a cycle. This is all discussed in May's book [23, Chap. 4]. In the Appendix he discusses a particular model in detail and gives a diagram [23, p. 192, Fig. A.1] showing the values of the parameters for which the equilibrium is stable or unstable. Changing the parameters so as to cross from the stable to the unstable region is an example of a Hopf bifurcation.

Both sorts of bifurcations are discovered by looking at the linear approximation of the differential equation near the equilibrium --essentially the derivative of the vectorfield at the equilibrium point. If the eigenvalues of the linear part are all negative or have negative real part if complex, then the equilibrium is asymptotically stable. The first sort of bifurcation is caused by a real eigenvalue changing from negative to positive as the parameter of the equation changes. The Hopf bifurcation occurs when a complex conjugate pair of eigenvalues cross the imaginary axis--i.e. the real part changes from negative to positive. The important point here is that complex eigenvalues are needed for a Hopf bifurcation. In general, the real part of an eigenvalue tells the radial rate at which the solution approaches or leaves the equilibrium. The imaginary part tells the angular rate at which the solution spirals around in some plane containing the equilibrium. So the imaginary part is needed to give the twist around the origin which becomes a cycle when the radial rate changes signs.

Hopf bifurcations can't occur in families of gradient fields because—as mentioned above—cycles can't occur. In fact, complex eigenvalues can't occur. In essence, the matrix of the linearization of the gradient of f is $(\partial^2 f/\partial x_i \partial x_j)$ and this is symmetric. Symmetric matrices have only real eigenvalues.

Now in Chap. IV we define for a vectorfield X on $\overset{\circ}{\Delta}$ the Hessian $H_p X$ at each point p of X. It is a real-valued bilinear map on the tangent space $T_p \overset{\circ}{\Delta}$ closely related to the linearization of the vectorfield. This Hessian is symmetric at every point p of $\overset{\circ}{\Delta}$ if and only if the vectorfield X is a gradient with respect to the Shahshahani metric. Since recombination is not a gradient there exist points p of $\overset{\circ}{\Delta}$ where the Hessian is not symmetric. Fix such a point p. By adding in various symmetric terms we can get any real parts we want for the eigenvalues in particular negative, zero and positive. Since the Hessian is not symmetric we can also make sure that some of the eigenvalues have nonzero imaginary parts. The variation of the symmetric part of the Hessian comes from the selection field.

We saw with equation (6.6) that we could think of the fitness numbers m_{ij} as built from $\bar{m}$, $m_i - \bar{m}$ and θ_{ij}. Now with p fixed we construct the selection fields to do what we want. First, choose $\bar{m}$ at p to be anything, say 0. Second, choose $m_i - \bar{m}$ at p to equal $-X_i(p)$ where $\Sigma X_i(p)\partial_i$ is the field we are looking at. This means that whatever θ_{ij} is, $\bar{\nabla}(\frac{1}{2}\bar{m}) + X$ has p as an equilibrium. Finally, θ_{ij} is an arbitrary symmetric matrix subject only to the conditions $\theta_i = \Sigma p_j \theta_{ij} = 0$ for all i. It turns out that there is still enough room to choose so that the θ_{ij}'s can be chosen to give any symmetric

Hessian $H_p(\bar{\triangledown}(\frac{1}{2}\bar{m}))$ we want on the subspace $T_p\mathring{\Delta}$ of R^I. So by varying our choice of θ's we can cause a Hopf bifurcation to occur at p.

The result is the most unsatisfying mathematical theorem: an existence proof. It leaves open several technical questions: Can the cycles which occur be asymptotically stable, i.e. limit cycles, like our simple examples? Are the cycles structurally stable or are they transient phenomena occurring only for peculiar isolated selection values? Answers to these questions require the construction and detailed study of particular examples. Also awaiting explicit examples is the more important question of the biological meaning of these examples. Do they have real significance or are they located in some biologically grotesque region of the space of selection matrices? In short, can the Wright Conjecture be revived for some restricted class of examples broader than zero-epistasis cases?

I mean no disrespect to Sewall Wright by attaching his name to a false conjecture. I thought it was true when I named it. Furthermore, as the above remarks indicate it may yet be true for real biological models. However, now I no longer think so.

I think of this result as analogous to Smale's result in [30]. There he shows that despite an apparently restrictive collection of axioms for equations modelling ecological competition, essentially any sort of dynamical behavior is possible given enough species. Here we are dealing with a much tighter class of models and within it we have discovered only the simplest sort of pathology, namely cycles. However, cycles are the first thing to look for after equilibria. Since we discovered them so easily, I suspect that much more complicated dynamical behavior lurks in these multilocus models. But again whether the dynamical complexity in Smale's models or ours is biological or merely mathematical can only be elucidated by the study of particular examples.

II. The Geometry of Epistasis

1. Orthogonal Decompositions.

Let I be a set containing n elements. In Sec. 4 of the previous chapter we defined the subsets $P = \{x \in R^I : x_i \geq 0 \text{ for } i \in I\}$ and $\Delta = \{x \in P : \Sigma \, x_i = 1\}$. By the formula I.(4.1) we defined the Shahshahani metric, a Riemannian metric on $\overset{\circ}{P} = \{x \in R^I : x_i > 0 \text{ for } i \in I\}$. It restricts to a metric on $\overset{\circ}{\Delta} = \Delta \cap \overset{\circ}{P}$. In Table 4 of Sec. I.4 we defined the maps $E^a(x) = \Sigma \, a_i x_i$ and $L^b(x) = \Sigma \, b_i \, \ell n \, x_i$ for $x \in \overset{\circ}{P}$, with $a, b \in R^I$. Note that the associations $a \to E^a$ and $b \to L^b$ are linear.

Now let A, B be a splitting of R^I orthogonal with respect to the usual inner product, $(\ ,\ )$. That is, R^I is the direct sum of subspaces A and B and $a \in A$, $b \in B$ imply $\Sigma \, a_i b_i = (a, b) = 0$. Choose bases $\{a^1, \ldots, a^t\}$ for A and $\{b^1, \ldots, b^d\}$ for B, so that $d + t = n$. Define the maps $E^A : R^I \to R^t$ and $L^B : \overset{\circ}{P} \to R^d$ by

$$E^A(x) = (E^{a^1}(x), \ldots, E^{a^t}(x))$$

$$L^B(x) = (L^{b^1}(x), \ldots, L^{b^d}(x)).$$

The above notation is somewhat abusive in that E^A and L^B really depend on the choice of bases, but a different choice of basis just changes the definition by multiplication by a nonsingular matrix in the range. Actually one can define an invariant version of E^A and L^B mapping to the dual spaces of A and B respectively so that the choice of bases simply amounts to choosing coordinates on the range. $E^A : \overset{\circ}{P} \to A^*$ and $L^B : \overset{\circ}{P} \to B^*$ would then be defined by:

$$\langle E^A(x), a \rangle = E^a(x) = (a, x).$$

(1.1)

$$\langle L^B(x), b \rangle = L^b(x) = (b, \ell n \ x),$$

where for $x \in \overset{\circ}{P}$ we denote by $\ell n \ x$ the vector whose i^{th} coordinate is $\ell n \ x_i$. Thus, $\ell n : \overset{\circ}{P} \to R^I$ is a diffeomorphism.

$E^A : R^I \to R^t$ is an onto linear map with kernel B and L^B is the composition of ℓn with $E^B : R^I \to R^d$, an onto linear map with kernel A. Thus, E^A maps $\overset{\circ}{P}$ onto an open cone $\overset{\circ}{P}_A$ in R^t and P onto P_A (the closure of $\overset{\circ}{P}_A$) while L^B maps $\overset{\circ}{P}$ onto all of R^d. Furthermore, it is clear that

$$E^A(x_1) = E^A(x_2) \quad \text{iff} \quad x_1 - x_2 \in B$$

(1.2)

$$L^B(x_1) = L^B(x_2) \quad \text{iff} \quad \ell n \ x_1 - \ell n \ x_2 \in A.$$

<u>1 Theorem</u>: Let $A \oplus B = R^I$ be an orthogonal splitting with respect to the usual inner product $(\ ,\ )$ on R^I, with $1 \in A$. Let $\{a^\alpha : \alpha = 1, \ldots, t\}$ and $\{b^\beta : \beta = 1, \ldots, d\}$ be bases for A and B respectively. Define $E^A : R^I \to R^t$ and $L^B : \overset{\circ}{P} \to R^d$ by

$$E^A(x)_\alpha = E^{a^\alpha}(x) = (a^\alpha, x) = \sum_i a_i^\alpha x_i$$

(1.3)

$$L^B(x)_\beta = L^{b^\beta}(x) = (b^\beta, \ell n \ x) = \sum_i b_i^\beta \ell n \ x_i,$$

and let $\overset{\circ}{P}_A = E^A(\overset{\circ}{P})$. $\overset{\circ}{P}_A$ is an open cone in R^t and $E^A \times L^B : \overset{\circ}{P} \to \overset{\circ}{P}_A \times R^d$ is a diffeomorphism with the following diagram commuting:

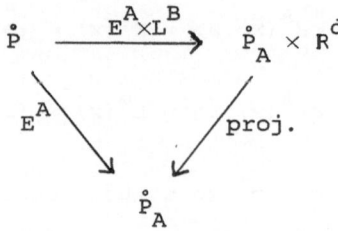

We will denote by $\mathfrak{D}$ the "foliation of fibres" of E_A whose
leaves are of the form $(E^A)^{-1}$ (point). The leaves of $\mathfrak{D}$ are bounded,
convex sets each open in a hypersurface of dimension d and L^B yields
a diffeomorphism of each leaf with R^d. We will denote by $\mathcal{J}$ the
"transverse foliation" whose leaves are of the form $(L^B)^{-1}$ (point).
With respect to the Shahshahani metric these foliations yield an
orthogonal splitting at each point. For $x \in \overset{\circ}{P}$

(1.4)
$$T_x\mathfrak{D} = \{\nabla_x L^b : b \in B\} = [\nabla_x L^{b^\beta} : \beta = 1,\ldots,d]$$

$$T_x\mathcal{J} = \{\nabla_x E^a : a \in A\} = [\nabla_x E^{a^\alpha} : \alpha = 1,\ldots,t]$$

where $[\ldots]$ means the subspace spanned by the listed vectors.

<u>Proof:</u> By I (4.13) the subspaces of $T_x P$: $\{\nabla_x E^a\}$ and $\{\nabla_x L^b\}$ are
perpendicular with respect to $(\ ,\ )_x$. Since they have bases $\{\nabla_x E^{a^\alpha}\}$
and $\{\nabla_x L^{b^\beta}\}$ respectively and since $d + t = n = \dim T_x P$, they form an
orthogonal splitting of $T_x P$ with respect to $(\ ,\ )_x$.

$E^A \times L^B$ maps $\overset{\circ}{P}$ to R^{d+t} and the gradients of the components
are everywhere linearly independent. So the derivative of this map
is an isomorphism at every point. It follows from the inverse function theorem that $E^A \times L^B$ is locally a diffeomorphism. Furthermore
the tangent space of $\mathfrak{D}$ at x is the $(\ ,\ )_x$ perpendicular complement
of $[\nabla_x E^{a^\alpha} : \alpha = 1,\ldots,t]$ which is the space spanned by the $\nabla_x L^b$'s.

(1.4) follows.

To complete the proof we have to show that $E^A \times L^B : \overset{\bullet}{P} \to \overset{\bullet}{P}_A \times R^d$ is one-to-one and onto. Since E^A maps $\overset{\bullet}{P}$ onto $\overset{\bullet}{P}_A$ it suffices to choose $x^0 \in \overset{\bullet}{P}$, let $z^0 = E^A(x^0)$ and show that L^B maps $(E^A|\overset{\bullet}{P})^{-1}(z^0) = (E^A)^{-1}(z^0) \cap \overset{\bullet}{P} = (x^0 + B) \cap \overset{\bullet}{P}$ one-to-one and onto R^d. Because $1 \in A$, the set $(x^0 + B) \cap \overset{\bullet}{P} \subset \{x: |x| = |x^0|\} \cap \overset{\bullet}{P}$ and so is bounded. We will prove bijectivity using the following topological lemma whose proof we will postpone until the appendix.

2 Lemma: Let U be an open set in Euclidean space and $F: U \to R^K$ be a local homeomorphism. If F is topologically proper, i.e. $F^{-1}(C)$ is compact whenever C is compact, then F is a homeomorphism and so is one-to-one and onto R^K.

So we are left with checking that on $(x^0 + B) \cap \overset{\bullet}{P} = \mathfrak{D}_{x^0} L^B$ is topologically proper. If it is not then there exists a sequence $\{x^n\}$ in $\mathfrak{D}_{x^0}$ with no subsequence convergent (in $\mathfrak{D}_{x^0}$) but with $L^B(x^n)$ bounded. Since $(x^0 + B) \cap P$ is compact we can assume by going to a subsequence that x^n converges to a point $x^\infty \in (x^0 + B) \cap (P - \overset{\bullet}{P})$. So $x_i^\infty \geq 0$ for all i, but the set $I_\infty = \{i: x_i^\infty = 0\}$ is nonempty. Let $b = x^\infty - x^0$. Then $b \in B$, and since $x^0 \in \overset{\bullet}{P}$, $b_i < 0$ for $i \in I_\infty$.

$$L^b(x^n) = \sum \{b_i \, \ell n \, x_i^n : i \in I_\infty\} + \sum \{b_i \, \ell n \, x_i^n : i \in I - I_\infty\}.$$

In the second sum each term is bounded as $n \to \infty$. In the first sum each term tends to $+\infty$ as $n \to \infty$. Hence, $L^b(x^n)$ tends to $+\infty$. But since b is a linear combination of the b^β's and $\{L^B(x^n)\}$ was assumed to be bounded, $\{L^b(x^n)\}$ is bounded. This contradiction completes the proof.

QED

<u>Remarks:</u> 1. In choosing the basis for A, we can assume that $a^t = 1$
and so $P_A \cap \{z \in R^t: z_t = 1\}$ $= \Delta_A$ and $\mathring{P}_A \cap \{z \in R^t: z_t = 1\} = \mathring{\Delta}_A$
are the images $E_A(\Delta)$ and $E_A(\mathring{\Delta})$ respectively. Δ_A is a closed convex
cell of dimension t with interior $\mathring{\Delta}_A$ (relative to $\{z_t = 1\}$).
$E^A \times L^B$ on $\mathring{P}$ restricts to a diffeomorphism in the diagram

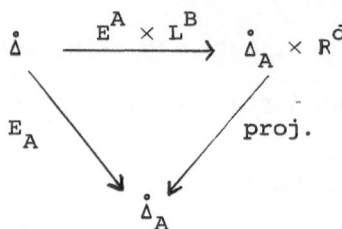

Restricting to $\mathring{\Delta}$, the foliation of fibres of E_A consists of those
leaves of $\mathfrak{D}$ which meet $\mathring{\Delta}$. The transverse foliation $\bar{\mathcal{J}}$ consists
of the leaves of $\mathcal{J}$ intersected with $\mathring{\Delta}$. For $p \in \mathring{\Delta}$

$$T_p\bar{\mathfrak{D}} = \{\bar{\nabla}_p L^b = \nabla_p L^b: b \in B\} = [\bar{\nabla}_p L^{b^\beta}: \beta = 1,\ldots,d]$$

(1.5)

$$T_p\bar{\mathcal{J}} = T_p(\mathcal{J} \cap \mathring{\Delta}) = \{\bar{\nabla}_p E^a: a \in A\} = [\bar{\nabla}_p E^{a^\alpha}: \alpha = 1,\ldots,t-1].$$

For the second equation we assume $a^t = 1$ and recall that $\bar{\nabla}_p E^1 = 0$.

2. In applications we will often define E^A mapping to R^ℓ where
$a^1,\ldots,a^\ell$ ($\ell \geq t$) is some spanning set for A. In this case $\mathring{P}_A$ is a
cone, open in a t-dimensional subspace, $E^A(R^I)$, of R^ℓ and by the
<u>Image Problem</u> we will denote the problem of describing this subspace.

(1.4), (1.5) and Table I.4.4 imply:

<u>3 Addendum:</u> (a) The vector $\Sigma\, x_i \xi_i \partial_i \in T_x \mathcal{J}$ iff $\xi \in A$ iff $\Sigma\, b_i \xi_i = 0$
for all $b \in B$. If $x = p \in \mathring{\Delta}$ then $\Sigma\, p_i \xi_i \partial_i \in T_p\bar{\mathcal{J}}$ iff in addition
$\bar{\xi} = \Sigma\, p_i \xi_i = 0$.

(b) The vector $\Sigma X_i \partial_i \in T_x \mathfrak{D}$ iff $X \in B$ iff $\Sigma a_i X_i = 0$ for all $a \in A$.

Recall that $\bar{\xi} = \Sigma p_i \xi_i = (\Sigma p_i \xi_i \partial_i, \nabla_p E^1)_p$. Note that in considering the transverse direction in part (a) we consider the relative rates ξ_i, while for the fibre direction in (b) we use the absolute rates X_i.

<u>4 Corollary</u>: Let f be a continuously differentiable real-valued function on P. The following are equivalent:

(a) f factors through E^A, i.e. there is a differentiable real-valued function $f^A: P_A \to R$ such that $f = f^A \cdot E^A$.

(b) For all $x \in P$, $\dfrac{\partial f}{\partial x} \in A$.

(c) For all $x \in \overset{\circ}{P}$, $(\nabla_x f, \nabla_x L^b)_x = 0$ for all $b \in B$.

<u>Proof</u>: By continuity the (unique) continuous factoring of f on P exists iff it exists on $\overset{\circ}{P}$. Since $\dfrac{\partial f}{\partial x_i}$ is continuous in x, (b) holds for all x iff it holds for $x \in \overset{\circ}{P}$. (b) on $\overset{\circ}{P}$ and (c) on $\overset{\circ}{P}$ are clearly equivalent and these are equivalent to f being constant on the fibres of E^A (see Prop. I.3.3), which is equivalent to the existence of a continuous factoring. Such a factoring is automatically as smooth as f because the onto linear map E^A has a right inverse $\bar{E}: R^t \to R^I$ and so $f^A = f \cdot \bar{E}$ on P_A.　　　　QED

<u>Remark</u>: The same result goes through if f is defined only on Δ and yields a factoring to a map on Δ_A. Simply apply the result to $f \cdot p$ where $p: P \to \Delta$ is the projection. Also since ∇L^b is perpendicular to p and ∇f differs from $\bar{\nabla} f$ by a multiple of p we can replace ∇f by $\bar{\nabla} f$ in (c).

For applications to fitness, the following special case is important:

5 Corollary: Let m be a symmetric element of $R^{I \times I}$. $\bar{m}: \Delta \to R$ is defined by $\bar{m}(p) = \Sigma \ p_i p_j m_{ij}$. The following are equivalent:

(a) $\bar{m}$ factors through E^A.

(b) For every $p \in \Delta$, the element of R^I defined by

$i \to m_i(p) = \Sigma \ p_j m_{ij}$ lies in A.

(c) For every $j \in I$, the element of R^I defined by $i \to m_{ij}$ lies in A.

(d) B is contained in the annihilator of the bilinear map defined by the matrix m.

Proof: Since $\dfrac{\partial \bar{m}}{\partial x_i}$ at p is equal to twice $m_i(p)$ the equivalence of (a) and (b) follows from Cor. 4. (b) implies (c) by using distribution p concentrated on j, i.e. $p_i = \delta_{ij}$ ($= 0$ or 1 as $i \neq j$ or $i = j$) and (c) implies (b) because A is closed under linear combination. Finally, (d) means that $\Sigma_i \ b_i m_{ij} = 0$ for all j and all $b \in B$. This is equivalent to (c) because A and B are orthogonal complements.   QED

For the foliation $\bar{\mathscr{J}}$ of $\mathring{\Delta}$ there is one particular distinguished leaf:

(1.6) $\Lambda_A = (L^B)^{-1}(0) = \{p \in \mathring{\Delta} : L^b(p) = 0 \text{ for all } b \in B\}$.

Λ_A is distinguished in being the leaf of greatest randomness relative to a choice of projection under E^A. Here randomness is measured by the entropy function $H(p) = -\Sigma \ p_i \ln p_i$. The meaning of this statement is the following:

6 Theorem: For $z \in \overset{\circ}{\Delta}_A$ let $\pi(z)$ be the point of Λ_A with $E^A(p(z)) = z$, i.e. let $\pi(z) = (E^A \times L^B)^{-1}(z,0)$. $\pi(z)$ is the unique point of $(E^A|\Delta)^{-1}(z)$ at which the entropy function H achieves its maximum. $\pi(z)$ is also the unique point p of $(E^A|\overset{\circ}{\Delta})^{-1}(z)$ such that ℓn p is a vector in A.

Proof: First note that ℓn $p \in A$ iff $L^b(p) = (b, \ell n\ p) = 0$ for all $b \in B$ iff $L^B(p)$ is the zero vector. Now define $N^B : \overset{\circ}{\Delta} \to R$ by $N^B(p) = \Sigma_\beta (L^{b_\beta}(p))^2$. $\bar{\nabla}N^B = 2 \Sigma_\beta L^{b_\beta}(p)\ \bar{\nabla}\ L^{b_\beta}(p)$. Thus, the gradient of N^B is tangent to the fibres of E^A, i.e. the leaves of $\overline{\mathcal{D}}$, by (1.5). By equation I(4.13):

$$(1.7) \qquad\qquad (\bar{\nabla}_p N^B, \bar{\nabla}_p H)_p = -2N^B(p).$$

Thus, as one moves down the gradient of N^B toward the point π where $N^B = 0$, or equivalently where L^B is the zero vector, H strictly increases.

To make this precise, define $\tilde{H} : \overset{\circ}{\Delta}_A \times R^+ \to R$ by:

$$\tilde{H}(z,s) = \sup\{H(p): E^A(p) = z \text{ and } N^B(p) = s\}.$$

Since $(E^A \times N^B)^{-1}(z,s)$ is compact (it is a d-1 dimensional sphere if $s > 0$ and a point if $s = 0$), $\tilde{H}$ is continuous. I claim that it is strictly decreasing in s. For let $s_1 > s_2 > 0$ and let $p_1 \in (E^A \times N^B)^{-1}(z,s_1)$ with $H(p_1) = \tilde{H}(z,s_1)$. Flow along the vector-field $-\nabla N^B$ starting at p_1 until one reaches a point p_2 with $N^B(p_2) = s_2$. Since the flow remains in the fibre of E^A, $E^A(p_2) = z$. Hence, $\tilde{H}(z,s_2) \geq H(p_2) > H(p_1) = \tilde{H}(z,s_1)$. It is clear that $\pi(z) = (E^A \times L^B)^{-1}(z,0)$ is the maximum of H on $(E^A|\overset{\circ}{\Delta})^{-1}(z)$. Furthermore, for any $s > 0$, $\{N^B > s\}$ is a neighborhood of the boundary in

$(E^A|_\Delta)^{-1}(z)$ and so it follows that $H|(E^{A-1}(z)) \cap (\Delta - \dot\Delta) \le$

$\inf\{\tilde{H}(z,): s \ge 0\} < \tilde{H}(z,0) = H((z))$. Thus $\pi(z)$ is the unique maximum point even when the boundary is included. QED

Remark: If $p \in \Lambda_A$ the the gradient of entropy at p, $\bar\nabla_p H$, is tangent to Λ_A, i.e. lies in $T_p\Lambda_A$. This is clear by direct computation and Addendum 2(a). Alternatively, since $H|\mathcal{D}_p$ achieves it maximum at p, $\bar\nabla_p H$ must be perpendicular to $T_p\mathcal{D}_p$ and so lies in $T_p\Lambda_A$ (since $\Lambda_A = \mathcal{I}_p$).

All of the results of this chapter can be reinterpreted in statistical language. The leaves of $\mathcal{I}$ correspond to what statisticians call log-linear restrictions on frequency tables for the finite set I. Such a restricted set of frequencies is $\{x \in \dot P: \Sigma\, b_i \ln x_i = K_b$ for all $b \in B\}$ where K_b are constants, depending on $b \in B$. This set is a leaf of $\mathcal{I}$ and different consistent choices of constants define different leaves. One of the consequences of Thm. 1 is that the only consistency condition needed to define a leaf is linearity of K_b in B.

Alternatively, if $x^0 \in \dot P$ then by (1.2) the leaf $\mathcal{I}_{x^0} = \{x \in \dot P: \ln x - \ln x^0 \in A\}$ i.e. $\mathcal{I}_{x^0}$ can be parametrized by A via the map

$$(1.8) \qquad\qquad x(a)_i = e^{a_i} x_i^0$$

$a \to x(a)$ defines a diffeomorphism of A onto $\mathcal{I}_{x^0}$ mapping 0 to x^0.

The leaves of $\bar{\mathcal{I}}$ in $\dot\Delta$ satisfy the additional restriction $\Sigma\, p_i = 1$. If $p^0 \in \dot\Delta$ then $\bar{\mathcal{I}}_{p^0}$ can be parametrized by A via the map:

$$(1.9) \qquad p(a)_i = C(a)^{-1} e^{a_i} p_i^0 \quad, \quad C(a) = \sum p_i^0 e^{a_i}.$$

The map $a \to p(a)$ is onto but not injective. $p(a^1) = p(a^2)$ iff the ratios $\exp(a_i^1) : \exp(a_i^2)$ are independent of i, i.e. iff $a^1 - a^2$ is a multiple of the vector 1. Thus, if we restrict the map to a complement A_0 of [1] in A we do get a diffeomorphism of A_0 onto $\bar{\mathcal{J}}_{p}^0$.

To a statistician, (1.9) says that each leaf of $\bar{\mathcal{J}}$ is a t-1 dimensional <u>exponential</u> <u>family</u> of distributions on the finite set I. Any exponential family of distribution on I can be exhibited this way.

In particular, since the leaf Λ_A clearly contains the center of the simplex p^0 with $p_i^0 = 1/n$, Λ_A can be parametrized by A_0 via:

(1.10)
$$p(a)_i = C(a)^{-1} e^{a_i} \qquad C(a) = \sum e^{a_i},$$

where we are absorbing the constant $1/n$ into C.

The theory of contingency tables [12] provides another viewpoint. The linear map E^A on Δ corresponds to what Gokhale and Kullback call the <u>design</u> <u>matrix</u>. In applications instead of knowing the entire distribution vector p we only know $z = E^A(p)$. The family $(E^A)^{-1}(z)$ of all distributions corresponding to z is, at least in $\mathring{\Delta}$, a leaf of the foliation $\bar{\mathfrak{D}}$. Of special interest in this leaf is the point $\pi(z)$ in Λ_A which is in some sense the distribution with the most independence among the elements of I subject to the constraint imposed by the design matrix and the fixed vector z. Now for $p \in \mathring{\Delta}$ let π be the unique element of $\Lambda_A \cap \bar{\mathfrak{D}}_p$, i.e. $E^A(p) = E^A(\pi)$ and $\pi \in \Lambda_A$. So $E^A \times L^B(\pi) = (E^A(p), 0)$. Define the normalized entropy:

(1.11) $$\hat{H}(p) = H(p) - H(\pi)$$

$$= -\sum p_i \ln p_i + \sum \pi_i \ln \pi_i$$

$$= -\sum p_i \ln(p_i/\pi_i).$$

The last equation in (1.11) is true because $p - \pi \in B$ by (1.1) and $\ln \pi \in A$ by Thm. 6. So with the usual inner product $(p, \ln \pi) = (\pi, \ln \pi)$.

7 Lemma: $\hat{H}: \mathring{\Delta} \to R$ satisfies the following:

    (a) $\hat{H}(p) \leq 0$ for $p \in \mathring{\Delta}$.

    (b) $\hat{H}(p) = 0$ iff $p \in \Lambda_A$.

    (c) With respect to the decomposition $T_p\mathring{\Delta} = T_p\bar{\mathcal{J}} \oplus T_p\bar{\mathfrak{D}}$ the $T_p\bar{\mathfrak{D}}$ components of $\bar{\nabla}_p H$ and $\bar{\nabla}_p \hat{H}$ agree.

    (d) $\bar{\nabla}_p \hat{H} = 0$ iff $p \in \Lambda_A$.

Proof: The function $H(\pi) = H - \hat{H}$ is constant on the leaves of $\bar{\mathfrak{D}}$. So the gradient of this function is everywhere perpendicular to $T_p\bar{\mathfrak{D}}$. This proves (c). On the fibre $(E^A)^{-1}(z)$ H has a strict maximum at π by Thm. 6. So $\hat{H}$ has a strict maximum at π and at π, $p = \pi$ so $\hat{H} = 0$. This proves (a) and (b). Now if $p \notin \Lambda_A$ then by (1.7)

$$(\bar{\nabla}_p N^B, \bar{\nabla}_p \hat{H})_p = (\bar{\nabla}_p N^B, \bar{\nabla}_p H)_p = -2N^B(p) \neq 0.$$

The two dot products agree by (c) because $\bar{\nabla}_p N^B \in T_p\bar{\mathfrak{D}}$. Finally, if $p \in \Lambda_A$, $\nabla_p H \in T_p\Lambda_A$ by the remark after Thm. 6. $\bar{\nabla}_p(H - \hat{H}) \in T_p\Lambda_A$ because $H - \hat{H}$ is constant on the leaves of $\bar{\mathfrak{D}}$. So $\bar{\nabla}_p \hat{H} \in T_p\Lambda_A$ for all $p \in \Lambda_A$. This means $(\bar{\nabla}_p \hat{H})$ on Λ_A is $\bar{\nabla}_p(\hat{H}|\Lambda_A)$ where this gradient is taken with respect to the Riemannian metric restricted to Λ_A. But

$\hat{H}|\Lambda_A$ is constantly zero. So its gradient is zero. This proves (d).

QED

On each fibre of $\bar{\mathfrak{D}}$ the negative $-\hat{H}$ is Kullback's <u>information</u> <u>discrimination</u> $I(p;\pi) = \Sigma\, p_i\, \ell n(p_i/\pi_i)$ see [12] and [21]. So the leaf Λ_A consists of the mimimum discrimination information (MDI) estimate of p subject to the design matrix constraint $E^A(p) = z$.

In the usual applications the set I is a product and the design matrix constraints correspond to knowing certain marginal distributions or joint distributions on some subproducts. This is exactly the case to which we now turn.

2. The Product Model.

The set I is the Cartesian product of the sets I_α, $\alpha = 1,\ldots,\ell$. $L = \{1,\ldots,\ell\}$ is the index set of loci. A complex K is a nonempty collection of subsets of L such that if $S_2 \subset S_1$ and $S_1 \in K$ then $S_2 \in K$. We will repeatedly identify a subset S of L with the complex consisting of S and all of its subsets. Note that the empty set, $\emptyset$, lies in every complex K. If $S \subseteq L\ I_S = \Pi\{I_\alpha: a \in S\}$ and if $i \in I$, i_S is the element of I_S whose α coordinate is i_α for all α in S. In this section we will often write $\xi(i)$ for ξ_i to avoid complicated subscripts.

We define $\mathscr{L}_K$ to be the subspace of R^I whose members are sums of functions depending only on blocs of loci in K. Thus

(2.1) $\mathscr{L}_S = \{\xi \in R^I: \text{ for some } \varphi \in R^{I_S}, \xi(i) = \varphi(i_S) \text{ for all } i \in I\}$.

(2.2) $\mathscr{L}_K = \displaystyle\sum \{\mathscr{L}_S: S \in K\}$.

Thus, $\xi \in \mathcal{L}_S$ if it depends only on the loci in S and $\xi \in \mathcal{L}_K$ if there exists $\varphi^S \in R^{I_S}$ for all $S \in K$ such that

(2.3) $$\xi(i) = \sum \{\varphi^S(i_S) : S \in K\}.$$

In particular, $\mathcal{L}_\emptyset$ consists of the constant functions.

It is clear that if K_1 and K_2 are complexes then the union $K_1 \vee K_2$ is and (2.2) implies:

(2.4) $$\mathcal{L}_{K_1 \vee K_2} = \mathcal{L}_{K_1} + \mathcal{L}_{K_2}.$$

If $i,j \in I$ and $S \subset L$, $i_S j_{\tilde{S}}$ denotes the element of I whose α coordinate is i_α for $\alpha \in S$ and j_α for $\alpha \in \tilde{S} = L - S$. Fix $j \in I$. Clearly, $\xi \in \mathcal{L}_S$ iff $\xi(i) = \xi(i_S j_{\tilde{S}})$ for all i since this means that the $\tilde{S}$ coordinates are irrelevant to the value of ξ. So we are led to define the following linear maps $P_S^j : R^I \to R^I$ and $D_S^j = 1 - P_S^j : R^I \to R^I$ by:

(2.5) $P_S^j(\xi)(i) = \xi(i_S j_{\tilde{S}})$, $D_S^j(\xi)(i) = \xi(i) - \xi(i_S j_{\tilde{S}})$ for $i \in I$.

Clearly, $(P_S^j)^2 = P_S^j$. P_S^j is a projection and D_S^j is the complementary projection. We recall some of the elementary properties of projections (see Halmos, [13, Sec. 41]).

<u>1 Lemma</u>: (a) If P is a projection on a vector space V then $\text{Im } P = \text{Ker } 1 - P = \{\xi \in V : P\xi = \xi\}$ and $\text{Ker } P = \text{Im } 1 - P = \{\xi \in V : P\xi = 0\}$. V is the direct sum of $\text{Ker } P$ and $\text{Im } P$.

(b) If P_1 and P_2 are projections which commute $(P_1 P_2 = P_2 P_1)$ then $P_1 P_2$ is a projection commuting with both and

(2.6a)
$$\text{Ker } P_1 P_2 = P_2^{-1}(\text{Ker } P_1) = \text{Ker } P_1 + \text{Ker } P_2$$

(2.6b)
$$\text{Im } P_1 P_2 = P_1(\text{Im } P_2) = (\text{Im } P_1) \cap (\text{Im } P_2)$$

(2.6c)
$$P_2^{-1}(\text{Im } P_1) = \text{Im } P_1 + \text{Ker } P_2$$

(2.6d)
$$P_1(\text{Ker } P_2) = (\text{Im } P_1) \cap (\text{Ker } P_2)$$

(c) If $\{P_i\}$ and $\{P_j\}$ are two finite families of projections all commuting with one another then:

(2.7)
$$(1 - \Pi_i P_i)(1 - \Pi_j P_j) = 1 - \Pi_{i,j}(1 - (1 - P_i)(1 - P_j))$$

If $V_i = \text{Ker } P_i$, $U_i = \text{Im } P_i$ and similarly for V_j and U_j, then:

(2.8a)
$$(\Sigma_i \ V_i) \cap (\Sigma_j \ V_j) = \Sigma_{i,j}(V_i \cap V_j)$$

(2.8b)
$$(\cap_i \ U_i) + (\cap_j \ U_j) = \cap_{i,j}(U_i + U_j).$$

Proof: (a): $(I - P)^2 = 1 - P$ and $P(1 - P) = (1 - P)P = 0$. Finally, if $\xi \in V$, $\xi = P\xi + (1 - P)\xi$ writes ξ uniquely as a sum from the image and the kernel of P.

(b): $(P_1 P_2)^2 = P_1 P_2$, $\text{Ker } P_1 P_2 \supset \text{Ker } P_2$ and $\text{Im}(P_1 P_2) \subset \text{Im } P_1$. Since $P_1 P_2 = P_2 P_1$, $\text{Ker } P_1 P_2 \supset \text{Ker } P_1$ and $\text{Im}(P_1 P_2) \subset \text{Im } P_2$. This proves half of (2.6a) and (2.6b). The other direction follows from:

$$1 - P_1 P_2 = (1 - P_1) + (1 - P_2) - (1 - P_1)(1 - P_2).$$

So if $\xi \in \text{Ker } P_1 P_2$, $\xi = (1 - P_1)\xi + (1 - P_2)\xi - (1 - P_1)(1 - P_2)\xi$ and so is in $\text{Im}(1 - P_1) + \text{Im}(1 - P_2) = \text{Ker } P_1 + \text{Ker } P_2$. If $\xi \in (\text{Im } P_1) \cap (\text{Im } P_2) = \text{Ker}(1 - P_1) \cap \text{Ker}(1 - P_2)$ then $(1 - P_1 P_2)\xi = 0$

and so $\xi \in \mathrm{Im}\ P_1 P_2$.

(2.6c) follows from (2.6a) applied to $1 - P_1$ and P_2. Similarly (2.6d) follows from (2.6b).

(c): We first note that if P_0 commutes with the family $\{P_i\}$ then

$$\Pi_i (P_0 + (1 - P_0)P_i) = P_0 + (1 - P_0)\Pi_i P_i.$$

For if we expand the product on the left we get P_0 and $(1 - P_0)\Pi_i P_i$ as end terms with all of the cross terms divisible by $P_0(1 - P_0) = 0$. Now we apply this equation twice:

$$\Pi_i P_i + (1 - \Pi_i P_i)\Pi_j P_j = \Pi_j(\Pi_j P_i + (1 - \Pi_i P_i)P_j)$$

$$= \Pi_j(P_j + (1 - P_j)\Pi_i P_i) = \Pi_{i,j}(P_j + (1 - P_j)P_i)$$

$$= \Pi_{i,j}(1 - (1 - P_i)(1 - P_j)).$$

This proves (2.7). (2.8a) follows by taking the image of both sides and (2.8b) follows by taking kernels. The equations are derived using (2.6a) and (2.6b). QED

<u>2 Proposition</u>: Fix $j \in I$.

(a) As S varies over the subsets of L the projections P_S^j and D_S^j all commute with one another.

(b) For K a complex define $D_K^j = \Pi\{D_S^j : S \in K\}$. D_K^j is a projection with kernel equal to $\mathcal{L}_K$. We let P_K^j denote the complementary projection $1 - D_K^j$.

(c) For complexes K_1 and K_2, the intersection $K_1 \wedge K_2$ is a complex and:

(2.9) $$\mathcal{L}_{K_1 \wedge K_2} = \mathcal{L}_{K_1} \cap \mathcal{L}_{K_2}.$$

<u>Proof</u>: (a): Since $D_S^j = 1 - P_S^j$ it suffices to check commutativity among the P_S^j's. Here it follows from:

$$P_{S_1 \cap S_2}^j = P_{S_1}^j \, P_{S_2}^j.$$

(b): We saw above that $\mathcal{L}_S$ is the Kernel of D_S^j. The general result now follows from (2.2) and (2.6a).

(c): We prove the following identities:

(2.10a) $$D_{K_1}^j \, D_{K_2}^j = D_{K_1 \vee K_2}^j$$

(2.10b) $$P_{K_1}^j \, P_{K_2}^j = P_{K_1 \wedge K_2}^j.$$

The first is clear from the definition of the projections D_K^j. Lemma 1(c) implies that with S_1 varying over K_1 and S_2 over K_2:

$$P_{K_1}^j \, P_{K_2}^j = [1 - \Pi_{S_1}(1-P_{S_1}^j)][1 - \Pi_{S_2}(1-P_{S_2}^j)] = 1 - \Pi_{S_1, S_2}(1-P_{S_1}^j \, P_{S_2}^j).$$

By the identity in the proof of (a) this is $1 - \Pi_{S_1, S_2}(1 - P_{S_1 \cap S_2}^j)$. This is $P_{K_1 \wedge K_2}^j$ because as S_1 and S_2 vary, $S_1 \cap S_2$ varies over the sets of $K_1 \wedge K_2$.

(2.9) follows from (2.10b) by taking Images and applying (2.6b).

QED

Thus, P_K^j for each j is a projection to $\mathcal{L}_K$. In applications we usually want projections which are orthogonal with respect to the covariance metric associated to a distribution p, i.e. to $_p(\ ,\ )$ (see Prop. I.4.3). In the special case when the different loci are independent we can get this projection from the P_K^j's and they have nice properties:

3 Proposition: For any $p \in \Delta$, define $D_K^p = \Sigma \, p_j D_K^j$ and $P_K^p = \Sigma \, p_j P_K^j = 1 - D_K^p$. D_K^p and P_K^p are complementary projections with $\text{Im } P_K^p = \text{Ker } D_K^p = \mathscr{L}_K$. If K_1 is any other complex

(2.11a)
$$P_K^p(\mathscr{L}_{K_1}) = \mathscr{L}_{K \wedge K_1} \quad \text{and} \quad P_K^{p-1}(\mathscr{L}_{K_1}) = \mathscr{L}_{K \vee K_1} .$$

(2.11b)
$$D_K^p(\mathscr{L}_{K_1}) \subseteq \mathscr{L}_{K_1 \setminus K} \qquad (K_1 \setminus K = \{S : S \subseteq S_1 \in K_1 - K\}) .$$

For $K = \{\emptyset\}$, $P_\emptyset^p (\xi)$ is the mean $\bar{\xi} = \Sigma \, p_j \xi_j$.

Now assume that p is a member of the Wright manifold Λ, i.e. $p_i = \Pi_\alpha p_{i_\alpha}^\alpha$ where p^α is the marginal distribution on the factor I_α induced by the distribution p. In that case D_K^p and P_K^p are orthogonal projections with respect to the covariance inner product $_p(\, , \,)$ on R^I. Furthermore, the following identities hold:

(2.12a)
$$D_{K_1}^p \cdot D_{K_2}^p = D_{K_1 \vee K_2}^p$$

(2.12b)
$$P_{K_1}^p \cdot P_{K_2}^p = P_{K_1 \wedge K_2}^p .$$

In particular, for fixed $p \in \Lambda$ the projections D_K^p and P_K^p all commute.

If p is the distribution concentrated at j, i.e. $p_i = \delta_{ij}$, then p is in Λ and P_K^p, D_K^p are the original projections P_K^j, D_K^j.

Proof: For all j, P_K^j maps R^I into $\mathscr{L}_K$ and is the identity on $\mathscr{L}_K$. These properties are preserved in forming the convex combination P_K^p and characterize a projection with image $\mathscr{L}_K$. The same convex combination argument allows us to get (2.11a) from the corresponding results for P_K^j. These results hold for P_K^j by (2.10a) and (2.10b) and the identities of Lemma 1. If $\xi \in \mathcal{L}_{K_1}$ then $\xi = \Sigma\{\varphi^S \colon S \in K_1\}$ with $\varphi^S \in \mathcal{L}_S$. $D_K^P(\varphi^S) = 0$ for $S \in K \wedge K_1$ and $D_K^P(\varphi^S) \in \mathcal{L}_S$ for the remaining S. So $D_K^P(\xi) = \Sigma\{D_K^P(\varphi^S) \colon S \in K_1 - K\} \in \Sigma\{\mathcal{L}_S \colon S \in K_1 - K\} = \mathcal{L}_{K_1 \setminus K}$. Since $P_\emptyset^j(\xi)$ is the constant $\xi(j)$, $P_\emptyset^P(\xi)$ is the constant $\Sigma \, p_j \xi_j = \bar{\xi}$.

The proof of the identities (2.12) hinge on the special case of (2.12b) where $K_1 = S_1$ and $K_2 = S_2$

$$P_{S_1}^P \cdot P_{S_2}^P (\xi)(i) = P_{S_1}^P \left(\sum p_j \xi(i_{S_2} j_{\widetilde{S}_2}) \right) = \sum p_j p_k \xi(I_{S_1 \cap S_2} k_{S_2 \cap \widetilde{S}_1} j_{\widetilde{S}_2}).$$

Summing first k over the variables not in $S_2 \cap \widetilde{S}_1$ and j over the variables in S_2 we get (letting, for example, $p(j_{\widetilde{S}_2})$ stand for the probability of $j_{\widetilde{S}_2}$):

$$= \sum p(j_{\widetilde{S}_2}) p(k_{S_2 \cap \widetilde{S}_1}) \xi(i_{S_1 \cap S_2} k_{S_2 \cap \widetilde{S}_1} j_{\widetilde{S}_2}).$$

Now because the different loci are independent $p(j_{\widetilde{S}_2}) p(k_{S_2 \cap \widetilde{S}_1}) = p(j_{\widetilde{S}_2} k_{S_2 \cap \widetilde{S}_1})$ and as $j_{\widetilde{S}_2}$ and $k_{S_2 \cap \widetilde{S}_1}$ range over $I_{\widetilde{S}_2}$ and $I_{S_2 \cap \widetilde{S}_1}$ the combined variable which we will call $r_{\widetilde{S}_1 \cup \widetilde{S}_2}$ ranges over $I_{\widetilde{S}_1 \cup \widetilde{S}_2}$. So we continue the chain of equations:

$$= \sum p(r_{\widetilde{S}_1 \cup \widetilde{S}_2}) \xi(i_{S_1 \cap S_2} r_{\widetilde{S}_1 \cup \widetilde{S}_2}) = P_{S_1 \cap S_2}^P (\xi)(i).$$

By induction it easily follows for a finite collection of sets $\{S_\mu\}$ with intersection S that

(2.13)
$$\Pi_\mu P_{S_\mu}^P = P_S^P = \sum p_j \Pi_\mu P_{S_\mu}^j.$$

In other words the average of the product is the product of the averages. Now D_K^j is $\Pi\{1 - P_S^j \colon S \in K\}$. Expand out the product, average over j, apply (2.13) and pack the product back together and we get

(2.14)
$$D_K^p = \Pi\{1 - P_S^p : S \in K\}.$$

From this (2.12a) is clear. We can rewrite (2.12b) as

$$(1 - D_{K_1}^p) \cdot (1 - D_{K_2}^p) = 1 - D_{K_1 \wedge K_2}^p .$$

This is true when p = j by (2.10b). Expanding out the product, averaging over j and applying (2.12a) we get that it holds for all p.

Next we show that P_S^p is self-adjoint with respect to $_p(\ ,\ )$.
Recall that a projection P is an orthogonal projection if it is
self-adjoint because if $\xi \in$ Im P and $\eta \in$ Ker P then $(\xi,\eta) = (P\xi,\eta)$
$= (\xi,P\eta) = 0$ and so Ker P is the orthogonal complement of Im P (see
Halmos, [13, Sec. 75]). Since any algebraic combination of commuting
self-adjoint operators is again self-adjoint it follows that D_K^p and
P_K^p are self-adjoint and so are orthogonal projections. Now if
$\xi, \eta \in R^I$, then

$$_p(P_S^p\xi,\eta) = \sum p(i)p(j)\xi(i_S j_{\tilde{S}})\eta(i) = \sum p(i)p(j_{\tilde{S}})\xi(i_S j_{\tilde{S}})\eta(i).$$

Since the loci are independent $p(i) = p(i_S)p(i_{\tilde{S}})$ and so

$$_p(P_S^p\xi,\eta) = \sum p(i_S)p(i_{\tilde{S}})p(j_{\tilde{S}})\xi(i_S j_{\tilde{S}})\eta(i_S i_{\tilde{S}}),$$

summed over the variables $i_S \in I_S$, $i_{\tilde{S}} \in I_{\tilde{S}}$ and $j_{\tilde{S}} \in I_{\tilde{S}}$. The expression
on the right is symmetric in ξ and η and so $_p(P_S^p\xi,\eta) = {}_p(P_S^p\eta,\xi) = $
$_p(\xi,P_S^p\eta).$ QED

4 Corollary. Let n be the number of elements in I. The map
$P_K = n^{-1} \Sigma_j P_K^j$ is the projection of R^I on $\mathcal{L}_K$ orthogonal with respect to

the usual inner product $(\ ,\ )$ on R^I.

Proof: Let n_α be the cardinality of I_α. Then $n = \Pi_\alpha n_\alpha$ and the uniform probability distribution $p_i = n^{-1}$ has independent loci with marginal probabilities $p_{i_\alpha}^\alpha = n_\alpha^{-1}$. So Prop. 3 applies. With the uniform probability distribution p, $_p(\ ,\ ) = n^{-1}(\ ,\ )$ and so orthogonality with respect to these two inner product is the same.   QED

5 Definition: (a) For $\xi \in R^I$ the carrier of ξ, $K(\xi)$, is the smallest complex K such that $\xi \in \mathcal{L}_K$. Equivalently (by (2.9)) $K(\xi) = \wedge\{K: \xi \in \mathcal{L}_K\}$. More generally, if T is a subset of R^I the carrier of T, $K(T)$, is the smallest complex K such that $T \subset \mathcal{L}_K$. So $K(T) = \wedge\{K: T \subset \mathcal{L}_K\} = \vee\{K(\xi): \xi \in T\}$. If m is a function with values in R^I then the carrier of m, $K(m)$, is the carrier of the image of m.

(b) For $S \subset L$ the dimension of S, dim S, is the number of points of S minus 1. For a complex K the dimension, dim K, is the maximum dim S for $S \in K$. For $\xi \in R^I$ or m a function to R^I, dim ξ or dim m is the dimension of the carrier of ξ or m, respectively.

We defined the map $E^S: R^I \to R_S \equiv R^{I_S}$ (cf. I. (0.2)):

(2.15) $$E^S(x)(i_S) = \sum\{x(i_S j_{\widetilde{S}}): j_{\widetilde{S}} \in I_{\widetilde{S}}\}.$$

So if $x = p \in \Delta$, i.e. p is a distribution on I, $E^S(p)$ is the distribution on I_S induced by the projection from I to I_S. For example, if $S = \alpha$ $E^\alpha(p)$ is the marginal distribution on the factor I_α. In these cases, we will write $p(i_S)$ or $p^S(i_S)$ for $E^S(p)(i_S)$.

Now define for a complex K

$$R_K = \Pi\{R_S : S \in K\}$$

$$E_K = \Pi\{E^S : S \in K\} : R^I \longrightarrow R_K.$$

Note as a particular convention $R_\emptyset = R$ with $E^\emptyset(x) = |x| = \Sigma\, x_i$.

We are now going to apply the results of section 1 to the orthogonal decomposition with $A = \mathcal{L}_K$. Our earlier projection results enable us to analyze B. With $i,j \in I$ we inductively define $b_{ij}^K \in R^I$, using the Kronecker delta notation ($\delta_{ij} = 0$ if $i \neq j$ and $= 1$ if $i = j$):

(2.16)

$$b_{ij}^\emptyset(k) = \delta_{ik} - \delta_{jk}$$

$$b_{i,j}^{K\vee S} = b_{i,j}^K - b_{\bar{i},j}^K \qquad (\bar{i} = i_S j_{\widetilde{S}}).$$

6 Lemma: Let $p \in \overset{\circ}{\Delta}$, and $X, \xi \in R^I$ with $X_k = p_k \xi_k$ for all $k \in I$. Let $b = b_{i,j}^K$. Then

$$D_K^j(\xi)(i) = (b,\xi) = (\triangledown_p L^b, X)_p.$$

Proof: The second equation follows from the definition of the Shahshahani metric and the computation of $\triangledown L^b$ in Table I.4.4. The first equation is clear for $K = \emptyset$ and then follows by induction on the number of elements of K since:

$$D_{K\cup S}^j(\xi)(i) = (D_S^j D_K^j)(\xi)(i) = D_K^j(\xi)(i) - D_K^j(\xi)(i_S j_{\widetilde{S}})$$

$$= (b_{i,j}^K, \xi) - (b_{\bar{i},j}^K, \xi) = (b_{i,j}^{K\vee S}, \xi).$$

QED

7 Theorem: (a) The image of $E^K : R^I \to R_K$ is the subspace V_K defined by the obvious consistency conditions. That is, we say that $\{x^S\} \in R_K$

lies in V_K if $S_1, S_2 \in K$ imply $E^{S_1 \cap S_2}(x^{S_1}) = E^{S_1 \cap S_2}(x^{S_2})$ where we get $E^{S_1 \cap S_2}(x^{S_1})$, for example, by summing on the variables of $S_1 - (S_1 \cap S_2)$. In particular, $|x^{S_1}| = |x^{S_2}|$ and this common value defines $|\{x^S\}|$ for $\{x^S\} \in V_K$.

(b) E^K maps $\overset{\circ}{\Delta}$ onto a convex set denoted $\overset{\circ}{\Delta}_K$ which is open in the hypersurface of V_K defined by $|\{x^S\}| = 1$. The closure of $\overset{\circ}{\Delta}_K$, denoted Δ_K, is the image of Δ.

(c) The kernel of E^K is the subspace B_K of R^I spanned by $\{b^K_{i,j} : i,j \in I\}$. Define $L^B: \overset{\circ}{\Delta} \to R^d$ by choosing a basis for B_K. Then the diagram

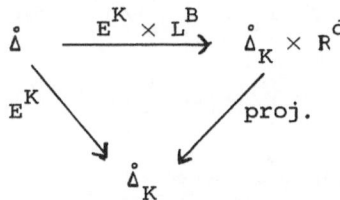

commutes and $E^K \times L^B$ is a diffeomorphism.

The fibres of E^K form the foliation $\overline{\mathfrak{D}}_K$ of $\overset{\circ}{\Delta}$ and the fibres of L^B form the transverse foliation $\overline{\mathcal{J}}_K$. $\overline{\mathfrak{D}}_K$ and $\overline{\mathcal{J}}_K$ are orthogonal with respect to the Shahshahani metric.

<u>Proof:</u> (a): We prove this in two parts.

(i) <u>When K consists of all proper subsets of L.</u> This case we prove by induction on the number of loci. We are given $\{y^\alpha(i_1 \ldots \hat{i}_\alpha \ldots i_\ell) : \alpha = 1, \ldots, \ell\} \in V_k$ where y^α is a function of all of the variables except i_α and we want to find $x \in R^I$ such that summing x in the α variable, we get:

$$x(i_1 \ldots *_\alpha \ldots i_\ell) = y^\alpha(i_1 \ldots \hat{i}_\alpha \ldots i_\ell).$$

Here an asterisk denotes summation over the indices of the labelled locus.

By inductive hypothesis we can define for each value of i_ℓ a function z^{i_ℓ} defined on $i_1 \ldots i_{\ell-1}$ which sums propertly for all α's except $\alpha = \ell$. So we define

$$\bar{y}^\alpha(i_1 \ldots \hat{i}_\alpha \ldots i_\ell) = y^\alpha(i_1 \ldots \hat{i}_\alpha \ldots i_\ell) - z(i_1 \ldots *_\alpha \ldots i_\ell).$$

Then $\bar{y}^\alpha$ is identically 0 for $\alpha = 1, \ldots, \ell-1$ and by the consistency conditions on $\{y^\alpha\}$,

$$\bar{y}^\ell(i_1 \ldots *_\alpha \ldots \hat{i}_\ell) = \bar{y}^\alpha(i_1 \ldots \hat{i}_\alpha \ldots *_\ell) = 0 \qquad (\alpha < \ell).$$

Now define $q(i_1 \ldots i_\ell) = \bar{y}^\ell(i_1 \ldots i_{\ell-1})/n_\ell$. Then

$$q(i_1 \ldots *_\alpha \ldots i_\ell) = \bar{y}^\alpha(i_1 \ldots \hat{i}_\alpha \ldots *_\ell)/n_\ell = 0 \qquad (\alpha < \ell),$$

by the above equation for $\bar{y}^\ell$. Also

$$q(i_1 \ldots i_{\ell-1} *_\ell) = \bar{y}^\ell(i_1 \ldots i_{\ell-1} \hat{i}_\ell).$$

Thus, $x = z + q$ is a solution with ℓ loci for $\{y^\alpha\}$.

(ii) <u>General Case</u>: If $K = L$ it is clear that V_K can be identified with R^I. So by adding subsets S one at a time it suffices to show that $V_{K \vee S}$ maps onto V_K when all proper subsets of S (the boundary of S) lie in K. So we are given a consistent family $\{x^T \in R_T : T \in K\}$ and want to find $x^S \in R_S$ consistent with this family. We can find $x^S \in R_S$ consistent with $\{x^T : T \subsetneq S\}$ by applying case (i) to the set of loci in S. But if T_1 is any element of K, $T = T_1 \cap S \subsetneq S$. x^{T_1} projects to x^T by consistency of the original family while x^S projects to x^T by consistency on S. Thus, x^S is consistent with the entire original family.

(b) - (c): These are direct applications of Thm. 1.1 and results which follow it. The proof is a matter of checking that with $A = \mathcal{L}_K$, B_K is the orthogonal complement of A and E^K is a version of E^A. First, Lemma 6 and Prop. 2(b) imply that B_K is the orthogonal complement of $\mathcal{L}_K$. The coordinate functions of the maps E^S are of the form E^a with $a = a_{k_S}$ where:

$$(2.17) \qquad a_{k_S}(i) = \delta_{i_S k_S} \qquad (k_S \in I_S).$$

Thus $a_{k_S}(i) = 1$ when i has specified values at the α positions for all $\alpha \in S$ and $= 0$ otherwise. Regarded as a function of i, a_{k_S} clearly depends only on i_S and so $a_{k_S} \in \mathcal{L}_S$. On the other hand, if $\varphi \in R_S$ and $\xi(i) = \varphi(i_S)$ then

$$\xi = \sum \{\varphi(k_S) a_{k_S} : k_S \in I_S\}.$$

Thus, the vectors $\{a_{k_S} : k_S \in I_S\}$ span $\mathcal{L}_S$ and the union of these sets with $S \in K$ spans $\mathcal{L}_K$. QED

Remark: The description of the subspace V_K of R_K by the linear consistency conditions in part (a) solves the Image Problem referred to after Thm. 1.1 for the case $A = \mathcal{L}_K$. This in turn leads to an obvious conjecture about the description of Δ_K. $\{p^S\} \in \Delta_K$ is a collection of probability distributions on the subsets S, satisfying the consistency conditions, i.e. p^{S_1} on I_{S_1} and p^{S_2} on I_{S_2} induce the same distribution $p^{S_1 \cap S_2}$ on the common subproduct $I_{S_1 \cap S_2}$. Is every such collection the image of a distribution p in Δ, i.e. does there exist a distribution p on I inducing the given p^S on the subproduct I_S for all S in K? Part (a) says that there exists $x \in R^I$

with $E^S(x) = p^S$ for all S and clearly $\Sigma_i \, x_i = |\{p^S\}| = 1$. It is

not clear from (a) that x can be chosen with $x_i \geq 0$ for all i. But

it does seem reasonable that the positivity of all of the p^S's should

allow us to choose some positive x, i.e. an element of Δ. Reason-

able, yes, but false. A counter example is given in the Appendix.

8 Theorem: Let $X(p) = \Sigma \, p_i \xi_i(p) \partial_i$ be a vectorfield on $\mathring{\Delta}$. So ξ_i is

a function of p for each i and $\bar{\xi} = \Sigma \, p_i \xi_i = 0$. Let K be a com-

plex of subsets of $L = \{1, \ldots, \ell\}$. The following conditions are

equivalent:

    (a)   The carrier of ξ is contained in K.

    (b)   For each $p \in \mathring{\Delta}$, $\xi(p) \in \mathscr{L}_K$.

    (c)   For all $S \in K$ there exist functions $\varphi^S : \mathring{\Delta} \to R^{I_S}$ as smooth

as ξ such that $\xi(i) = \Sigma\{\varphi^S(i_S) : S \in K\}$ at every point $p \in \mathring{\Delta}$.

    (d)   For all $i, j \in I$ and every point $p \in \mathring{\Delta}$ we have

$$\sum_k b^K_{i,j}(k)\, \xi(k) = 0.$$

    (e)   The vectorfield X is everywhere tangent to the trans-

verse foliation $\bar{\mathcal{J}}_K$.

    (f)   For every vector b in the vector space B_K spanned by

$\{b^K_{i,j} : i, j \in I\}$, the function L^b is an integral of the motion for the

vectorfield X.

    If X is the gradient of a function $f : \mathring{\Delta} \to R$, i.e. $X = \bar{\nabla} f$ or,

equivalently, $\xi_i = \dfrac{\partial f}{\partial x_i}$ at every point p of $\mathring{\Delta}$ then the above condi-

tions are further equivalent to:

    (g)   The function f factors (uniquely) through the map E^K.

Proof: Except for the smoothness in (c) the equivalence of (a) - (c) is a matter of definitions. They are equivalent to (d) because B_K is the orthogonal complement of $\mathcal{L}_K$ with respect to $(\ ,\ )$. (b) is equivalent to (e) by Addendum 1.3(a), (d) is equivalent to (f) because L^b is an integral of the motion iff the gradient of L^b is orthogonal to X at all points. In the gradient case equivalence with (g) follows from Cor. 1.4 and the Remark after it.

The smoothness portion of (c) is proved by induction on the number of subsets in K. If $K = S$, the projection from I to I_S induces an isomorphism from $R_S = R^{I_S}$ onto $\mathcal{L}_S$ and we define φ^S to be the composition of ξ with the inverse isomorphism (and $\varphi^{S_1} = 0$ for $S_1 \subsetneq S$). If $K = K_1 \vee S$ and P_{K_1}, D_{K_1} are defined as in Cor. 4, then $P_{K_1} \circ \xi$ maps into $\mathcal{L}_{K_1}$ and by (2.11b) $D_{K_1} \circ \xi$ maps to $\mathcal{L}_S$. Applying the initial step to the latter and the inductive hypothesis to the former we decompose $\xi = P_{K_1} \circ \xi + D_{K_1} \circ \xi$ as the sum of functions as smooth as ξ.

$$\text{QED}$$

Recall that the selection field is the gradient $\bar{\nabla}(\frac{1}{2}\bar{m})$ where $\bar{m} = \Sigma\, p_i p_j m_{ij}$ is mean zygotic fitness. m_{ij} can be regarded as an element of $R^{I \times I}$ where, in addition, m_{ij} is symmetric in i and j.

<u>9 Theorem</u>: Let $I = \Pi_{\alpha=1}^{\ell} I_\alpha$ and consider m_{ij} a symmetric function on $I \times I$ (i.e. $m_{ij} = m_{ji}$ for $i,j \in I$). Let K be a complex in L. The following are equivalent:

(a) On Δ, then function $\bar{m}(p) = \Sigma\, p_i p_j m_{ij}$ factors through E^K.

(b) $m_i(p) = \Sigma\, p_j m_{ij}$ is a function of p from Δ to $\mathcal{L}_K$.

(c) For every $S \in K_1$ there exists $\varphi^S \colon \Delta \to R^{I_S}$ such that $m_i(p) = \Sigma_{S\in K}\varphi^S(p)_{i_S}$.

(d) For every $j \in I$, $m_{ij} \in \mathcal{L}_K$ as a function of i.

(e) For all $S,T \in K$, there exists $m^{ST} \in R^{I_S \times I_T}$ such that $m_{ij} = \Sigma_{S,T} m^{ST}(i_S, j_T)$. Furthermore, we can assume that $m^{ST}(i_S, j_T) = m^{TS}(j_T, i_S)$ for all $S,T \in K$.

Recall that m_{ij} is _completely symmetric_ if whenever $Q \subset L$, $\bar{i} = i_Q j_{\tilde{Q}}$ and $\bar{j} = j_Q i_{\tilde{Q}}$ then $m_{ij} = m_{\bar{i}\bar{j}}$. If m_{ij} is completely symmetric we can strengthen (e) to:

(e_{sym}) For all $S \in K$ there exists $m^S \in R^{I_S \times I_S}$ completely symmetric in the variables of I_S such that $m_{ij} = \Sigma_S m^S(i_S, j_S)$.

Proof: The equivalence of (a) - (d) follows from Thm. 8 and Cor. 1.5. Clearly, (e) implies (d). On the other hand, (d) implies that for every $j \in I$ and $S \in K$ there exists $m_j^S \in R^{I_S}$ such that $m_{ij} = \Sigma\{m_j^S(i_S): S \in K\}$. Now consider $m_j^S(i_S)$ as a function of j, i.e. as an element of R^I with i_S fixed and project to K. For every $T \in K$ there exists $m^{ST} \in R^{I_S \times I_T}$ such that:

$$\sum \{m^{ST}(i_S, j_T): T \in K\} = P_K(m_j^S(i_S)).$$

Sum on S:

$$\sum \{m^{ST}(i_S, j_T): S,T \in K\} = P_K(m_{ij}) = m_{ij}.$$

The last equality because by symmetry and (d) $m_{ij} \in \mathcal{L}_K$ as a function of j. Finally, by symmetry:

$$m_{ij} = \frac{1}{2}(m_{ij} + m_{ji}) = \sum \{[\frac{1}{2}(m^{ST}(i_S, j_T) + m^{TS}(j_T, i_S))]: S,T \in K\}.$$

So we can replace $m^{ST}(i_S, j_T)$ by the bracketed term to get symmetry in (e).

The complete symmetry result is less direct. For $Q \subset L$, define

$T_Q(m_{ij}) = m_{\bar{i}\bar{j}}$ (where $\bar{i}$ and $\bar{j}$ are defined in the statement). T_Q is a linear operator on $R^{I \times I}$ with $(T_Q)^2 = $ identity.

The index set for $I \times I$ is two copies of L. We can indicate subsets of the doubleof L by ordered products $S \times T$ with $S, T \subset L$. If K_1 and K_2 are complexes on L then $K_1 \times K_2 = \{S \times T : S \in S_1$ and $T \in K_2\}$ is a complex on the double of L. T_Q acts on subsets of the double and hence on complexes by:

(2.18) $T_Q(S \times T) = [(S \cap Q) \cup (T \cap \tilde{Q})] \times [(T \cap Q) \cup (S \cap \tilde{Q})].$

Clearly, $T_Q^2 = $ identity on complexes, too. In the notation of Def. 5:

(2.19) $$T_Q(Car(m)) = Car(T_Q(m)).$$

That $Car(T_Q(m)) \subset T_Q Car(m)$ is clear from writing

(2.20) $$m(i,j) = \sum \{m^{ST}(i_S, j_T) : S \times T \in Car(m)\}$$

and applying T_Q to both sides. Note that $T_Q(m^{ST}) \in R^{I_{\bar{S}} \times I_{\bar{T}}}$ where $\bar{S} \times \bar{T} = T_Q(S \times T)$. On the other hand, T_Q is clearly monotone on complexes, i.e. $T_Q(K_1) \subset T_Q(K_2)$ if $K_1 \subset K_2$ as complexes on the double. Hence:

$$Car(m) = Car(T_Q^2 m) \subset T_Q \, Car(T_Q m) \subset T_Q^2 \, Car(m) = Car(m).$$

Since the extremes are equal these are all equal and so (2.19) holds after T_Q is applied. Since $T_Q^2 = $ identity, this implies (2.19).

Now (e) implies that $Car(m) \subset K \times K$. I claim that complete symmetry implies $Car(m) \subset 2K$ where:

(2.21) $$2K = \{S \times T : S \cup T \in K\}.$$

Complete symmetry means $T_Q(m) = m$ for all $Q \subset L$. Hence, $\mathrm{Car}(m)$ is invariant under T_Q by (2.19). If $S \times T \in \mathrm{Car}(m)$ and $Q = S$ then $T_Q(S \times T): (S \cup T) \times (S \cap T)$ which must be in $\mathrm{Car}(m) \subset K \times K$ and so $S \cup T \in K$. Now any function $m^{ST} \in R^{I_S \times I_T}$ is a. fortiori a function $m^U \in R^{I_U \times I_U}$ where $U = S \cup T$. Hence, the decomposition in (e_{Sym}) follows from (2.20). To get complete symmetry of the functions m^S we average as before: $m_{ij} = 2^{-\ell} \Sigma \{T_Q(m_{ij}): Q \subset L\}$. So we can replace m^S by $2^{-\ell} \Sigma \{T_Q(m^S): Q \subset L\}$. QED

<u>Remark</u>: The above proof shows that if m_{ij} is symmetric and for each j fixed we regard m_{ij} as a function of i, then $\mathrm{Car}_i(m_{ij}) \subset K$ for all j implies $\mathrm{Car}_{ij}(m_{ij}) \subset K \times K$. Furthermore, if m_{ij} is completely symmetric $\mathrm{Car}_{ij}(m_{ij}) \subset 2K$.

We saw in Prop. 3 that for p in the Wright manifold Λ, the projection of R^I onto $\mathcal{L}_K$ orthogonal with respect to the covariance metric $_p(\ ,\ )$ is given by P_K^p. For general p this projection is difficult to compute. However, we can interpret it geometrically and relate it to the concept of conditional expectation.

First, a technical result:

<u>10 Lemma</u>: Let X_1 and X_2 be Riemannian manifolds and $E: X_1 \to X_2$ be a submersion, i.e. the derivative map $d_p E: T_p X_1 \to T_{Ep} X_2$ is an onto linear map for all p in X_1. Let V_p (the vertical subspace) denote the kernel of $d_p E$ and H_p denote the perpendicular complement of V_p in $T_p X_1$. $d_p E$ is an isomorphism of H_p on $T_p X_2$. Following O'Neill [26], we call E a Riemannian submersion if $d_p E$ is an isometry of H_p on $T_p X_2$ for all p. E is a Riemannian submersion iff the following condition holds:

For every smooth map $f: X_2 \to R$, the gradient of $f \cdot E$ at $p \in X_1$ is the unique horizontal vector mapping under $d_p E$ to the gradient of f at Ep, i.e.

(2.22) $$d_p E(\nabla_p (f \cdot E)) = \nabla_{Ep} f.$$

Proof: Since $f \cdot E$ is constant on the fibres of E, the gradient of $f \cdot E$ is always horizontal, whether E is a Riemannian submersion or not. The question is whether equation (2.22) holds.

$d_p E$ is an isomorphism on H_p, so we can designate by $d_p E(v)$ an arbitrary vector in $T_{Ep} X_2$ with $v \in H_p$. Then:

$$(d_p E(v), d_p E(\nabla_p (f \cdot E)))_{Ep} \overset{(1)}{=} (v, \nabla_p (f \cdot E))_p$$

$$\overset{(2)}{=} d_p (f \cdot E)(v) = d_{Ep} f(d_p E(v)).$$

Equation (1) holds iff E is a Riemannian submersion while (2) is by definition of the gradient. Finally, the end terms are equal iff (2.22) holds. The equivalence of the lemma follows because any vector in $T_{Ep} X_2$ can be obtained as $\nabla_{Ep} f$ for some choice of smooth $f: X_2 \to R$.

QED

11 Proposition: Let $p \in \overset{\circ}{\Delta}$.

(a) Identify $\mathscr{L}_S$ with R_S by the isomorphism induced by the projection $I \to I_S$ for $S \subset L$. The projection of R^I on $\mathscr{L}_S$ which is orthogonal with respect to the covariance metric $_p(\ , \ )$ ($p \in \Delta$) is the conditional expectation operator $E_p(\ |S)$, that is, the value of the projection of ξ at i_S is:

(2.23) $$E_p(\xi | i_S) = \sum \{ p(j) \xi(j) / p(i_S) : j_S = i_S \}.$$

Define $d_i^S = p(i) - p(i_S)p(i_{\tilde{S}})$. This measure of the failure of independence of the S loci from the $\tilde{S}$ loci relates the orthogonal
projection with the projection P_S^p of Prop. 3. For $i_S \in I_S$:

$$(2.24) \qquad \sum \{d_j^S \xi(j) : j_S = i_S\} = p(i_S)[\mathbb{E}_p(\xi|i_S) - P_S^p(\xi)(i_S)].$$

The following special cases are of interest: (1) If $\xi \in \mathcal{L}_S$
then $\mathbb{E}_p(\xi|S) = P_S^p(\xi) = \xi$ and the left side of (2.24) is 0. (2) If
$\xi \in \mathcal{L}_{\tilde{S}}$ and $\bar{\xi}$ is 0 then $P_S^p(\xi) = 0$ and the left side of (2.24)
equals $p(i_S)\mathbb{E}_p(\xi|i_S)$. (3) If the loci of S and $\tilde{S}$ are indepen-
dent then $d_i^S = 0$ for all i and the left side of (2.24) is 0. So
in this case $\mathbb{E}_p(\xi|S) = P_S^p(\xi)$.

(b) The linear map $E^K: \mathring{\Delta} \to \mathring{\Delta}_K$ is a submersion with the vertical subspace at p equal to $T_p\overline{\mathfrak{D}}_K$. With respect to the Shahshahani
metric the horizontal subspace is equal to $T_p\bar{\mathcal{J}}_K$. In the notation of
Thm. 7 (a), we define $V_K^0 = \{\{x^S\} : |\{x^S\}| = 0\}$ and identify the tan-
gent space of $\mathring{\Delta}_K$ with the subspace V_K^0 at all points. The derivative
of the linear map E^K is E^K itself. E^K restricts to an isomorphism
of $T_p\bar{\mathcal{J}}_K$ on V_K^0. The composition $(E^K|T_p\bar{\mathcal{J}}_K)^{-1} \circ E^K$ is the projection of
$T_p\mathring{\Delta}$ on $T_p\bar{\mathcal{J}}_K$ orthogonal with respect to the Shahshahani metric. When
F_p is the isomorphism of Prop. I.4.3 the following diagram commutes:

$$
\begin{array}{ccc}
(\mathbb{R}^I)_0 & \xrightarrow{\;\;F_p\;\;} & T_p\mathring{\Delta} \\
\downarrow & & \downarrow \\
(\mathcal{L}_K)_0 & \xrightarrow{\;\;F_p\;\;} & T_p\bar{\mathcal{J}}_K
\end{array}
\qquad (p \in \mathring{\Delta}).
$$

Here $(\mathcal{L}_K)_0$ and $(\mathbb{R}^I)_0$ consist of the members of $\mathcal{L}_K$ and $\mathbb{R}^I$ respectively
with mean equal to zero. The left and right vertical maps are pro-
jections orthogonal with respect to the covariance and Shahshahani

metrics respectively. When $p \in \Lambda_0$ the left map is P_K^p (c.f. Prop. 3).

If $K = S$ then with respect to the Shahshahani metric on the simplex $\mathring{\Delta}_S$ too, $E^S : \mathring{\Delta} \to \mathring{\Delta}_S$ is a Riemannian submersion. The tangent map at p is given by the formula (for $X = \Sigma \, p(i)\xi(i)\partial_i$):

$$(2.25) \qquad E^S(X) = \sum \{ p(i_S)\mathbb{E}_p(\xi \mid i_S)\partial_{i_S} : i_S \in I_S \}.$$

(c) If $f : \mathring{\Delta} \to R$ is a smooth map then the orthogonal projection of the gradient $\bar{\nabla}_p f$ to $T_p \mathcal{J}_K$ is the same as the gradient of the restriction of f to the leaf of $\bar{\mathcal{J}}_K$ through p, $f \mid (\bar{\mathcal{J}}_K)_p$. For the latter gradient the restriction of the Shahshahani metric is used.

Proof: We begin with (b). Any onto linear map is a submersion and the identification of the vertical and horizontal subspaces follow from Thm. 7 (c). $(E^K \mid T_p \bar{\mathcal{J}}_K)^{-1} \cdot E^K$ maps $T_p \mathring{\Delta}$ onto $T_p \bar{\mathcal{J}}_K$ and is the identity on $T_p \bar{\mathcal{J}}_K$. So it is a projection on $T_p \bar{\mathcal{J}}_K$. Since its kernel is the vertical subspace which is orthogonal to $T_p \bar{\mathcal{J}}_K$ it is the orthogonal projection. Prop. I.4.3 says that F_p is an isometry and it clearly maps $(\mathcal{J}_K)_0$ onto $T_p \bar{\mathcal{J}}_K$. From this the commutative diagram follows.

When $K = S$, $E^K(\partial_i) = \partial_{i_S}$ for all $i \in I$ and so (2.25) follows from the definition of $\mathbb{E}_p$ (which is (2.23)) and linearity. X is a horizontal vector iff $\xi \in (\mathcal{L}_S)_0$ meaning ξ depends only in i_S and so $\mathbb{E}_p(\xi \mid i_S) = \xi(i)$. So if $\xi, \eta \in (\mathcal{L}_S)_0$ with $p \in \Delta$,

$$_p(\xi, \eta) = \sum p(i)\xi(i)\eta(i) = \sum p(i_S)\xi_p(\xi \mid i_S)\xi_p(\eta \mid i_S).$$

Via the isometry F_p this says that E^S is an isometry of $T_p \bar{\mathcal{J}}_S$ on $T_p \mathring{\Delta}_S$. So E^S is a Riemannian submersion.

For (a), That $E_p(\xi|S)$ is the orthogonal projection on $\mathcal{L}_S$ is a classical result about conditional expectation. But it also follows from the diagram in (b) by the equation (2.25). Equation (2.24) and the special cases are easy direct computations.

(c) is a special case of a general result about the gradients of restrictions to submanifolds. If g is the restriction to $(\bar{\mathcal{J}}_K)_p$ then the gradient is characterized by: (1) $\bar{\nabla}_p g \in T_p \bar{\mathcal{J}}_K$ and (2) $(\bar{\nabla}_p g, X)_p = d_p g(X) = d_p f(X)$ for all $X \in T_p \bar{\mathcal{J}}_K$. The orthogonal projection of $\bar{\nabla}_p f$ satisfies these conditions. QED

<u>12 Corollary</u>: Let $p \in \mathring{\Delta}$. If $\xi \in \mathcal{L}_K$ then ξ is completely determined by the set of conditional expectations $\{E_p(\xi|i_S): s \in K\}$.

<u>Proof</u>: By subtracting $\bar{\xi} = E(\xi)$ if necessary we can assume that the mean is zero. So $X = \sum p(i)\xi(i)\partial_i$ lies in the tangent space of Δ at p. $E^S(X) = X^S = \sum p(i_S)E_p(\xi|i_S)\partial_{i_S}$. If $\xi \in \mathcal{L}_K$ then $X \in T_p \bar{\Delta}_K$ and so is the image of $\{X^S: S \in K\}$ under the isomorphism $(E^K|T_p\bar{\Delta}_K)^{-1}$. QED

We can now see that if $p \in \mathring{\Delta} - \Lambda$ the orthogonal projection of R^I onto $\mathcal{L}_K$ need not satisfy (2.11a).

For example, suppose $\xi \in \mathcal{L}_{\tilde{S}}$ and consider the orthogonal projection of ξ to $\mathcal{L}_S$. (2.11a) would say that the image lies in $\mathcal{L}_{S \cap \tilde{S}} = \mathcal{L}_{\{\emptyset\}}$ which is the set of constant functions. If ξ began with mean 0 then (2.11a) would imply that the image has to be zero. But unless i_S and $i_{\tilde{S}}$ are independent with respect to p equation (2.24) and special case (2) show that the projection need not be zero.

We now prove a result which is useful in computing K-type epistasis approximations to the selection field.

First, another lemma about projections:

<u>13 Lemma</u>: Let V be a Euclidean vector space with metric (,).

(a) If $T: V \to V$ is an isometric isomorphism and V_0 is a T invariant subspace, then the orthogonal projection, P, of V on V_0 commutes with T.

(b) Let $V_1 \supset V_2$ be subspaces of V. If $P_1: V \to V_1$, $P_2: V \to V_2$ and $P_{21}: V_1 \to V_2$ are orthogonal projections, then $P_2 = P_{21} \cdot P_1$.

<u>Proof</u>: (a): Let $v \in V$ and $u \in V_0$. By invariance, TPv and $T^{-1}u$ lie in V_0. So, because T is isometric and P is orthogonal:

$$(Tv - TPv, u) = (T(v - Pv), TT^{-1}u)$$

$$= (v - Pv, T^{-1}u) = 0.$$

So $TPv \in V_0$ with Tv - TPv perpendicular to V_0. Thus, TPv is the orthogonal projection of Tv on V_0, i.e. TPv = PTv.

(b): $P_{21} \cdot P_1(v) \in V_2$ for $v \in V$. $v - P_{21} \cdot P_1(v) = (v - P_1(v))$ $+ (P_1(v) - P_{21} \cdot P_1(v))$ and so is perpendicular to V_2. Thus, $P_{21} \cdot P_1(v)$ is the orthogonal projection of v on V_2, i.e. $P_{21} \cdot P_1(v) = P_2(v)$.   QED

<u>14 Proposition</u>: (a) If $\bar{p}_{ij}$ is a distribution on $I \times I$ which is completely symmetric as a function of i and j and K is a complex on L, then the $_{\bar{p}}(,)$ orthogonal projection of $R^{I \times I}$ onto $\mathcal{L}_{2K}$ (c.f. (2.21)) commutes with T_Q for all $Q \subset L$. In particular, such a projection preserves complete symmetry. If $\bar{p}_{ij} = p_i p_j$ for some distribution p on I, then $\bar{p}$ is completely symmetric iff all loci are independent, i.e. iff $p \in \Lambda$. In that case all of the loci of $I \times I$ are independent with respect to $\bar{p}$.

When K = S, the projection to $\mathcal{L}_{2S} = \mathcal{L}_{S \times S}$ preserves complete symmetry when the S and $\tilde{S}$ loci are independent, i.e.

$$d_i^S = p(i) - p(i_S)p(i_{\tilde{S}}) = 0 \text{ for all } i.$$

(b) There is a natural identification isomorphism between $\mathcal{L}_{L \times \{\emptyset\}}$, consisting of functions of ij depending only on i, and R^I. The identification maps $\mathcal{L}_{K \times \{\emptyset\}}$ isomorphically onto $\mathcal{L}_K$. If $\bar{p}_{ij} = p_i p_j$ then the identification is an isometry of $_{\bar{p}}(\ , \ )$, with $_p(\ , \ )$.

(c) Let $\bar{p}_{ij} = p_i p_j$ and let K be a complex on L. The following diagram commutes with the arrows all orthogonal projections and the isomorphisms the identification map of (b):

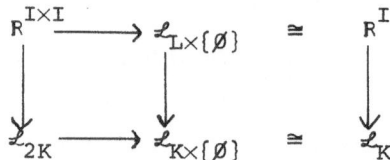

Proof: If $\bar{p}$ is completely symmetric then it is easy to check that T_Q is an isometry of $_{\bar{p}}(\ , \ )$. In fact, it is the map induced by a measure preserving bijection of $I \times I$. Since $T_Q(\mathcal{L}_{S \times T}) = \mathcal{L}_{\bar{S} \times \bar{T}}$ where $\bar{S} \times \bar{T} = T_Q(S \times T)$, it is clear that $\mathcal{L}_{2K}$ is T_Q invariant. That the projection commutes with T_Q and so preserves complete symmetry follows from Lemma 13 (a).

Clearly, if $p \in \Lambda$ and $\bar{p}_{ij} = p_i p_j$ then all of the loci are $\bar{p}$ independent and $\bar{p}$ is completely symmetric. For the converse, sum the equation $p(i)p(j) = p(i_Q j_{\tilde{Q}})p(i_{\tilde{Q}} j_Q)$ on j to get $p(i) = p(i_Q)p(i_{\tilde{Q}})$ or $d_i^Q = 0$. That this holds for all i and Q implies $p \in \Lambda$ by the proofs of Prop. I.5.1 and its Corollary.

Finally, if $m \in R^{I \times I}$ it is easy to check that

$$T_Q P^{k\ell}_{S\times S}(m)(ij) = m(i_{S\cap Q}j_{S\cap \widetilde{Q}}^{\,k}\widetilde{\,S}, j_{S\cap Q}i_{S\cap \widetilde{Q}}^{\,\ell}\widetilde{\,S}) = P^{k\ell}_{S\times S}T_{QU\widetilde{S}}(m)(ij).$$

Averaging $k\ell$ by the distribution $\bar{p}_{k\ell}$ we get for all $\bar{p}$:

(2.26)
$$T_Q P^{\bar{p}}_{S\times S} = P^{\bar{p}}_{S\times S}T_{QU\widetilde{S}}.$$

Now if $\bar{p}_{ij} = p_i p_j$ and $d^S_i = 0$ for all i then $d^{S\times S}_{ij} = 0$ for all ij in $I \times I$ and so by Prop. 11 (a) $P^{\bar{p}}_{S\times S}$ is the orthogonal projection. So that projection preserves complete symmetry.

(b): $(I \times I)_{L\times\{\emptyset\}}$ is just I and the identification is a special case of the identification between $\mathcal{L}_S$ and R_S by the projection $I \to I_S$. Preservation of $\mathcal{L}_{K\times\{\emptyset\}}$ is clear and the isometry result is an easy computation.

(c): The square on the left commutes because both compositions are the orthogonal projection to $\mathcal{L}_{K\times\{\emptyset\}}$ by Lemma 13 (b). The square on the right commutes because the identification in (b) is an isometry.

QED

Remark: By Prop. 11 (a) if $m \in R^{I\times I}$ and P is the $(\,,\,)$ projec-tion on $\mathcal{L}_{L\times\{\emptyset\}}$ followed by the identification with R^I:

$$P(m)_i = \mathbb{E}_{\bar{p}}(m|i) = \sum \{\bar{p}_{ij}m_{ij}/p_i : j \in I\}.$$

So if $\bar{p}_{ij} = p_i p_j$ the projection to R^I is $m_i = \Sigma\, p_i m_{ij}$. The diagram of Prop. 11 (c) then says that we can get the projection of m_i to $\mathcal{L}_K$ by first projecting m_{ij} to $\mathcal{L}_{2K}$ to get $n_{ij} \in \mathcal{L}_{2K}$ and then averaging n_{ij} to get $n_i \in \mathcal{L}_k$. If m_{ij} is completely symmetric and $p \in \Lambda$ then part (a) implies that n_{ij} is completely symmetric.

In applying these results to K type epistasis for a particular complex K we need a basis or at least a spanning set for the (,)

perpendicular complement B_K of the subspace $\mathcal{L}_K$. By Thm. 7 (c) the

set $\{b_{ij}^K : i \in I\}$ and any fixed choice of j is such a spanning set.

More generally, Prop. 3 and Lemma 6 imply that for any fixed $p \in \Delta$

$\{\Sigma_j \, p_j b_{ij}^K : i \in I\}$ spans B_K. However, special examples of complexes

K may have special spanning sets associated with them. We now

construct such a set for the case where K is the s <u>skeleton</u>

$L^{(s)}$ of L (cf. Sec. 2 of Chap. I):

(2.27) $L^{(s)} = \{S \subset L: \dim S \leq s\}.$

Recall from Def. 5 that the dimension of S is the cardinality of

S minus one. Thus, for example, $L^{(0)}$ consists of all of the

singleton sets $\{\alpha\}$ of L together with $\emptyset$. If $K = L^{(s)}$ we will write

$\mathcal{L}^{(s)}$ for $\mathcal{L}_K$.

For $L^{(s)}$ we get a useful spanning set by beginning, as in the

general case, with a linear operator.

For every order preserving map $\lambda: L \to \{0,\ldots,s+1\}$, and every

map $\epsilon: \{0,\ldots,s+1\} \to \{0,1\}$ and $i,j \in I$ define: $(i,j,\lambda,\epsilon) \in I$ by

(2.28) $(i,j,\lambda,\epsilon)_\alpha = \begin{cases} i_\alpha & \epsilon \circ \lambda(\alpha) = 0 \\ \\ j_\alpha & \epsilon \circ \lambda(\alpha) = 1. \end{cases}$

The map λ partitions L is s + 2 disjoint subsets namely

$\lambda^{-1}(q)$ for $q = 0,\ldots,s+1$. $\epsilon \circ \lambda$ then labels each subset with either a

0 or a 1. (i,j,λ,ϵ) is the member of I agreeing with i on the

subsets labelled with a 0 and with j on the subsets labelled with

1. So $(i,j,\lambda,\epsilon) = i_S j_{\tilde{S}}$ for $S = (\epsilon \circ \lambda)^{-1}(0) \subset L$.

Now define $|\epsilon| = \epsilon(0)+\ldots+\epsilon(s+1) = $ the number of times 1 is hit by ϵ, and for $\xi \in R^I$ define

(2.29)
$$D_{i,j}^{\lambda}(\xi) = \sum (-1)^{|\epsilon|} \xi(i,j,\lambda,\epsilon)$$

where the sum is taken over the 2^{s+2} maps ϵ.

As in Lemma 6, if $X_i = p_i \xi_i$ and $L_{i,j}^{\lambda}(p)$ is defined for $p \in \overset{\circ}{\Delta}$ by

(2.30)
$$L_{i,j}^{\lambda}(p) = \sum_{\epsilon} (-1)^{|\epsilon|} \ell n \; P_{(i,j,\lambda,\epsilon)}$$

and we have

(2.31)
$$(\nabla_p L_{i,j}^{\lambda}, X)_p = D_{i,j}^{\lambda}(\xi).$$

To relate $D_{i,j}^{\lambda}$ to the previous operators, define
$S(\lambda,q) = \overbrace{\lambda^{-1}(q)}^{} = L - \lambda^{-1}(q)$ and note that

$$D_{i,j}^{\lambda}(\xi) = (\prod_{q=1}^{s+1} D_{S(\lambda,q)}^j)(\xi)(i).$$

Here the proof is just a matter of expanding the product out using
$D_S^j = 1 - P_S^j$ and the equation (cf. (2.10b)):

$$\xi(i,j,\lambda,\epsilon) = \prod_q \{P_{S(\lambda,q)}^j : \epsilon(q) = 1\}(\xi)(i).$$

Thus, $D_{i,j}^{\lambda}(\xi) = 0$ for all i,j and λ iff for every λ, ξ is
the sum of terms each concentrated on some $S(\lambda,q)$ i.e. not depen-
dent on the variables in $\lambda^{-1}(q)$ for some q. Now if ξ depends on
some bloc of $s + 1$ variables λ applied to these variables misses
some $q \in \{0,\ldots,s+1\}$, i.e. the bloc is contained in some $S(\lambda,q)$.
Hence, if $\xi \in \mathcal{L}^{(s)}$ then $D_{i,j}^{\lambda}(\xi) = 0$ for all i,j and λ. On the other
hand if the carrier of ξ has dimension greater than S and so
contains some set S with dim $S \geq s + 1$ and upon which ξ depends
then choosing λ to that λ maps S onto $\{0,\ldots,s+1\}$, $S \nsubseteq S(\lambda,q)$

for any q, and so by definition of the carrier of ξ, ξ can't lie in $\mathscr{L}_K$ where $K = \vee_q S(\lambda,q)$. So some $D^\lambda_{i,j}(\xi)$ is not 0. Thus, $\mathscr{L}^{(s)}$ is the intersection of the kernels of D^λ_{ij}. This proves the following:

15 Lemma: Define $b^\lambda_{ij} \in R^I$ so that with $b = b^\lambda_{ij}$, $L^\lambda_{ij} = L^b$. That is, let $(b^\lambda_{ij})_k$ be the coefficient of $\ln p_k$ in (2.30). $\{b^\lambda_{ij} : i,j \in I$ and $\lambda : L \to \{0,\dots,s+1\}\}$ is a spanning set for $B^{(s)}$.

Remarks: 1. The above proof shows that we need only include surjective maps λ to get a spanning set. If λ is surjective and i,j are fixed in I then the 2^{s+2} expressions $(i,j,\lambda,\varepsilon)$ are all distinct. So from (2.30) the coefficient of $\ln p_k$ in $L^\lambda_{ij}(p)$ is either ± 1 or 0.

2. In the other direction, we did not need that λ was order-preserving to define D^λ_{ij} or to show $b^\lambda_{ij} \in B^{(s)}$. The order-preserving condition merely reduces the size of the spanning set.

III. Selection, Recombination and Mutation

1. Selection and Epistasis.

The general model of frequency dependent selection is a vector-field $X = \Sigma\, X_i \partial_i = \Sigma\, p_i \xi_i \partial_i$ on Δ where the components X_i and ξ_i are C^∞ functions of the state $p \in \Delta$. We think of $\xi_i(p)$ as the relative fitness of gametic genotype i when the population distribution is at p. The fitness is relative because ξ is normalized to mean zero, i.e. $\Sigma\, p_i \xi_i(p) = \Sigma\, X_i(p) = 0$. Restricting to the interior distributions $\mathring{\Delta}$, we interpret the results of Chap. II to describe at most K-type epistasis or K-type epistasis, for short, when K is a complex in the set of loci L.

Thm. II.2.8 gives a list of equivalent conditions which define what we mean by K-type epistasis. It means (by part (c) of the Thm.) that there exist C^∞ functions $\varphi^S : \mathring{\Delta} \to R^{I_S}$ for $S \in K$ such that

$$(1.1) \qquad\qquad \xi_i(p) = \sum_S \varphi^S_{i_S}(p).$$

So for each fixed $p \in \mathring{\Delta}$, $\xi_i(p)$ regarded as a function of i is a sum of terms each depending only on the alleles in some bloc S of loci in K, i.e. $\xi(p) \in \mathcal{L}_K$. Note that, in general, the coordinate function $\varphi^S_{i_S}$ depend on the entire distribution p and not just on the partial distribution p^S induced on I_S, i.e. they do not factor through E^K.

We can test for K-type epistasis using part (d): X has K-type epistasis if $\Sigma\, b_i \xi_i(p) = 0$ for all $p \in \mathring{\Delta}$ whenever b is a vector of the $(\ ,\ )$ orthogonal complement B_K of $\mathcal{L}_K$. In applying this test it suffices to check it for b in some spanning set for B_K.

Geometrically, K-type epistasis means that the leaves of the transverse foliation $\bar{\mathcal{J}}_K$ are invariant with respect to solutions of the differential equation associated to X. Since these leaves are defined by the equations $\{L^b = \text{constant}\}$ this is a conservation principle. The leaves act like energy levels with energy conserved by selection. To be precise, for $b \in B_K$, the function

$$(1.2) \qquad\qquad L^b(p) = \ln \prod_i (p_i)^{b_i}$$

remains constant as p changes under the influence of a selection field exhibiting K-type epistasis. Alternatively, the gene ratios which are the antilogs of the functions L^b remain constant for $b \in B_K$ iff the selection field satisfies K-type epistasis.

If X is the gradient vectorfield of some fitness function $f: \overset{\circ}{\Delta} \to R$, i.e. $X(p) = \bar{\triangledown}_p f$, then by part (g) X satisfies K-type epistasis iff f factors through the map E^K. This means that the value of f at p depends only on induced distributions p^S for $S \in K$. In contrast to the general case above, for a gradient field the functions φ^S do factor through E^K.

1 Proposition: Let X be a gradient vectorfield satisfying K-type epistasis. Each component ξ_i regarded as a function of p factors through E^K. Furthermore, the functions φ^S for $S \in K$ can be chosen to factor through E^K.

Proof: Recall from equation I. (4.12) that if $X = \bar{\triangledown} f$ then for any smooth extension of $f: \overset{\circ}{\Delta} \to R$ to a function on $\overset{\circ}{P}$, $\xi_i = \partial f/\partial x_i - \Sigma\, p_j\, \partial f/\partial x_j$. In particular we can use the extension defined by:

(1.3) $\qquad f(x) = f(x/|x|) \qquad x \in \overset{\circ}{P}.$

This is the unique extension of f to a function on $\overset{\circ}{P}$ which is homogeneous of degree zero. By Euler's Thm. $\Sigma\, x_j\, \partial f/\partial x_j = 0$ for such a function (see Shahshahani [28, p. 2]) and so we have

(1.4) $\qquad \xi_i(p) = \dfrac{\partial f}{\partial x_i}\Big|_p \quad$ for all $p \in \overset{\circ}{\Delta}.$

Now for $b \in B_K$ we know that $0 = \Sigma\, b_i \xi_i = \Sigma\, b_i(\partial_f/\partial x_i)$ at every point p of $\overset{\circ}{\Delta}$. By homogeneity this equation holds for the extension of f at all points of $\overset{\circ}{P}$. So we can differentiate again to get:

(1.5) $\qquad 0 = \displaystyle\sum_i b_i\, \frac{\partial^2 f}{\partial x_i \partial x_j} = \sum_i b_i\, \frac{\partial \xi_j}{\partial x_i} \qquad$ for all $b \in B_K.$

Applying Thm. II.2.8 to ξ_j with j fixed we have that $\overline{\nabla}_p \xi_j \in \mathcal{L}_K$ and so ξ_j factors through E^K. Now the inductive construction of the functions φ^S in the proof of Thm. II.2.8 yield functions all factoring through E^K. $\qquad\qquad$ QED

Remark: That X was gradient was really used in (1.5) which needs $\partial \xi_i/\partial x_j = \partial \xi_j/\partial x_i$ for all i and j.

Our selection field is a gradient with $f = \frac{1}{2}\,\overline{m}$. $\overline{m}$ is mean fitness defined by the symmetric matrix m_{ij} of fitness constants. Now Thm. II.2.9 applies to describe K type epistasis of the selection field. Furthermore, in the completely symmetric case the decomposition of the gametic fitness function into a sum of K-blocs extends to the zygotic fitness numbers (cf. (e_{sym})):

$$(1.6) \qquad m_{ij} = \sum_S m^S(i_S, j_S) \qquad m^S \in R^{I_S \times I_S}; \qquad S \in K.$$

To get m_i we multiply by p_j and sum on j. Apply this to the above equation and reverse the order of summation. Then for each fixed S sum on the complementary $\tilde{S}$, loci first. Applying equation II.(2.15) and letting p^S denote $E^S(p)$, we get:

$$(1.7) \qquad m_i = \sum_S m^S(i_S),$$

where

$$m^S(i_S) = \sum_{j_S} p^S(j_S) m^S(i_S, j_S).$$

Similarly:

$$(1.8) \qquad \bar{m} = \sum_S \overline{m^S}$$

where

$$\overline{m^S} = \sum_{i_S, j_S} p^S(i_S) p^S(j_S) m^S(i_S, j_S).$$

We now turn to some examples:

Zero Epistasis ($K = L^{(0)}$): This is the case described in detail in Sec. 5 of Chap. I. The projection E ($= E^{(0)}$ we omit the superscript for this case alone) is the product $\Pi_\alpha E^\alpha$ mapping $\mathring{\Delta}$ onto $\Pi_\alpha \mathring{\Delta}^\alpha$. $\xi \in R^I$ lies in $\mathcal{L}^{(0)}$ iff there exist $\varphi^\alpha \in R^{I_\alpha}$ ($\alpha = 1, \ldots, \ell$) with

$$(1.9) \qquad \xi_i = \sum_\alpha \varphi^\alpha_{i_\alpha} \qquad \text{for} \quad i \in I.$$

The orthogonal complement $B^{(0)}$ is spanned by the coefficient vectors of L^S_{ij} defined by I. (5.2). To see this apply Lemma II.2.15 and the remarks thereafter. Here s = 0 and so λ maps L to {0,1}. Letting S = λ^{-1}(0) the four elements of I of the form (i,j,λ,ϵ) corresponding to the four maps ϵ: {0,1} $\to$ {0,1} are just i,j,$\bar{\text{i}}$ = $i_S j_{\tilde{S}}$, and $\bar{\text{j}}$ = $j_S i_{\tilde{S}}$. So L^λ_{ij} = L^S_{ij} in this case. Choosing a basis from among these maps we define L: $\overset{\bullet}{\Delta} \to R^d$. In particular, the leaf of maximum entropy L^{-1}(0) is the Wright manifold $\overset{\bullet}{\Lambda}$ in this case.

Lemma II.2.15 shows that to get a spanning set we need only consider order preserving maps λ. Such a map is determined by a choice of $\mu \in L$ so that λ^{-1}(0) = S = {α $\quad \alpha \leq \mu$}. We adopt Shahshahani's notation and let (i:μ:j) be the element of I whose allele at the α loci is i_α if $\alpha \leq \mu$ and is j_α if $\alpha > \mu$. Then L^S_{ij} is:

(1.10)
$$L^\mu_{ij} = \ln[p_i p_j / p_{(i:\mu:j)} p_{(j:\mu:i)}]$$

$$= \ln p_i - \ln p_{(i:\mu:j)} - \ln p_{(j:\mu:i)} + \ln p_j.$$

Since the coefficient vectors of the L^μ_{ij}'s span $B^{(0)}$, $\xi \in \mathcal{L}^{(0)}$ if for all i,j $\in$ I and $\mu \in$ L:

(1.11)
$$\xi_i - \xi_{(i:\mu:j)} - \xi_{(j:\mu:i)} + \xi_j = 0.$$

The mathematical fact that the L^μ_{ij}'s span the same space as the larger set of L^S_{ij}'s is essentially the same as the biological fact that an exchange of genetic material between the chromosomes at exactly the loci in S for any subset S of L can be obtained by a sufficiently intricate (and so sufficiently unlikely) sequence of single cross overs.

If m_{ij} is completely symmetric then Thm. II.2.9 says that the following conditions are equivalent and define zero epistasis for the selection field $\bar{\triangledown}(\frac{1}{2}\bar{m})$:

(a) Mean fitness $\bar{m}$ factors through E i.e. depends only on the individual gene frequencies $p(i_\alpha)$.

(b) For all μ, i and j, L^μ_{ij} is an integral of the selection vectorfield $\bar{\triangledown}\frac{1}{2}\bar{m}$, i.e. the ratios $p_i p_j / P_{(i:\mu:j)} P_{(j:\mu:i)}$ remain the same as the genotype distribution changes only according to selection.

(c) The leaf $\overset{\circ}{\Lambda}$ is left invariant by selection, i.e. $\bar{\triangledown}\frac{1}{2}\bar{m}$ is tangent to $\overset{\circ}{\Lambda}$ at points of $\overset{\circ}{\Lambda}$.

(d) There exist symmetric functions $m^\alpha \in R^{I_\alpha \times I_\alpha}$ such that

$$(1.12) \qquad m_{ij} = \sum_\alpha m^\alpha_{i_\alpha j_\alpha}.$$

Some remark about (c) is in order. The other conditions imply that $\bar{\triangledown}\bar{m}$ is everywhere tangent to the transverse foliation $\bar{\mathcal{F}}$. This implies (c). Conversely, (c) means that $m_i(p)$ as a function of i lies in $\mathcal{L}^{(0)}$ for $p \in \overset{\circ}{\Lambda}$. Let p approach the distribution concentrated at j through distributions in $\overset{\circ}{\Lambda}$. $m_i(p)$ approaches m_{ij} and so m_{ij} lies in $\mathcal{L}^{(0)}$ as a function of i for all j. This is condition (d) of Thm. II.2.9.

As (1.12) is a special case of (1.6) we get the corresponding special cases of (1.7) and (1.8):

$$(1.13) \qquad m_i = \sum_\alpha m^\alpha_{i_\alpha} = \sum_\alpha p^\alpha_{j_\alpha} m^\alpha_{i_\alpha j_\alpha}.$$

$$(1.14) \qquad \bar{m} = \sum_\alpha \bar{m}^\alpha = \sum_\alpha p^\alpha_{i_\alpha} p^\alpha_{j_\alpha} m^\alpha_{i_\alpha j_\alpha}.$$

It is important to understand what these equations do not say.

$m_{i_\alpha}^\alpha$ need not be the average fitness of a gamete carrying allele i_α at the α locus. With $p \in \overset{\bullet}{\Delta}$ this average fitness is the conditional expectation of the random variable m_i assuming that the allele at the α locus is i_α. This expression is computed in Prop. II.2.11 (a):

$$(1.15) \qquad m(i_\alpha) = \mathbf{E}_p(m \mid i_\alpha) = m_{i_\alpha}^\alpha + \sum_{\beta \neq \alpha} \mathbf{E}(m^\beta \mid i_\alpha)$$

$$= m_{i_\alpha}^\alpha + \sum_{\beta \neq \alpha} \overline{m^\beta} + \sum_{\beta \neq \alpha} \sum_{i_\beta} d_i^{\alpha,\beta} m_{i_\beta}^\beta / p(i_\alpha)$$

where $d_i^{\alpha,\beta} = p(i_\alpha i_\beta) - p(i_\alpha)p(i_\beta)$. So $m(i_\alpha)$ depends on the joint distribution of i_α and i_β. Thus, even after normalizing to mean zero, $m(i_\alpha) - \bar{m}$ differs from $m_{i_\alpha}^\alpha - \overline{m}^\alpha$ by the third term in the sum which is due to linkage disequilibrium. Now if $p \in \overset{\bullet}{\Lambda}$ the loci are independent and so this last term does vanish. Furthermore on $\overset{\bullet}{\Lambda}$ we can describe the selection field quite simply:

<u>2 Proposition</u>: $E: \overset{\bullet}{\Lambda} \to \Pi_\alpha \overset{\bullet}{\Delta}_\alpha$ is a diffeomorphism. If we put the Shahshahani metric on $\overset{\bullet}{\Lambda}$ and the product of the Shahshahani metrics of $\overset{\bullet}{\Delta}_\alpha$ on the product Π, then E is an isometry of Riemannian manifolds.

If m is given by (1.12) then $\overline{\triangledown} \frac{1}{2} \bar{m}$ is tangent to $\overset{\bullet}{\Lambda}$ and maps under E to the vectorfield which on the factor $\overset{\bullet}{\Delta}_\alpha$ is $\overline{\triangledown} \frac{1}{2} \overline{m}^\alpha$ where the latter gradient is with respect to the Shahshahani metric on $\overset{\bullet}{\Delta}_\alpha$.

<u>Proof</u>: The isometry result is an easy computation using Prop. II. 2.11(b) and the orthogonality of vectors depending on i_α with vectors depending on i_β when $\alpha \neq \beta$ and $p \in \overset{\bullet}{\Lambda}$. The isometry result implies the gradient results which also follow by direct computation using Prop.

II.2.11(a) and (1.15)   QED

    This means that if there is zero epistasis the selection field

on $\overset{\circ}{\Lambda}$ is just the sum of the separate effects of the different loci

with each depending only on the gene frequencies at that locus. Thus,

there is here no problem of "genetic background".

    <u>One-dimensional Epistasis</u> $(K = L^{(1)})$:Again, apply Lemma II.

2.15. In this case λ is an order-preserving map onto $\{0,1,2\}$ and

so corresponds to picking two positions $1 \leq \mu < \nu < \ell$. L_{ij}^{λ} becomes

$$(1.16) \qquad L_{i,j}^{\mu,\nu} = \ell n \; \frac{P_i P_j P_{(i:\mu:j:\nu:i)} P_{(j:\mu:i:\nu:j)}}{P_{(i:\mu:j)} P_{(j:\mu:i)} P_{(i:\nu:j)} P_{(j:\nu:i)}} \; .$$

Thus, there is one-dimensional epistasis iff these ratios are pre-

served for all $\mu < \nu$, i and j. One dimensional epistasis is common,

implicitly, if not in the world at least in a lot of genetic models

because of the following:

<u>3 Proposition</u>: Let m_{ij} be given by (1.12) and let $F: R \rightarrow R$ be some

quadratic function, i.e. $F(t) = at^2 + bt + c$. If $n_{ij} = F(m_{ij})$ then

n_{ij} is completely symmetric and exhibits at most one-dimensional

epistasis.

<u>Proof</u>: Squaring out the sum one gets cross terms like $2am_{i_\alpha j_\alpha}^{\alpha} m_{i_\beta j_\beta}^{\beta}$

which depends on two loci. So n_{ij} is the sum of terms depending on

at most two loci each. The result follows from Thm. II.2.9. Complete

symmetry is obvious.   QED

    This quadratic model arises, for example, if we assume that the

genes act additively to yield some metric trait measured by m_{ij} and hat fitness n_{ij} is given by:

1.17)
$$n_{ij} = c_1 - c_2 (m_{ij} - m_0)^2$$

here m_0 is some optimal value of the trait.

Adjacent Locus Interaction ($K = \{1,2\} \vee \{2,3\} \vee \ldots \vee \{\ell-1,\ell\}$):
n this case we regard the set L of loci as lying in order along a
ingle chromosome. The only interactions are pairwise and of these
nly adjacent loci interact. So K is contained in $L^{(1)}$ but is
uch more restricted.

Proposition: For $K = \{1,2\} \vee \ldots \vee \{\ell-1,\ell\}$, a spanning set for B_K is
iven by the coefficient vectors of two families of function L_{ij}^{λ}:
$L_{ij}^{\mu,\nu}$: $\mu < \nu$ and $i,j \in I\}$ (cf. (1.16) and $\{L_{ij}^{\mu}$: $i,j \in I$ such that
$_{\mu} = j_{\mu}\}$ (cf. (1.10)).

roof: By the previous example, $(b,\xi) = 0$ for all $b = b_{ij}^{\mu,\nu}$ iff ξ
as only pairwise interactions, i.e. iff for all pairs $\{\alpha,\beta\}$ and $\alpha < \beta$
here exists $\varphi^{\alpha\beta} \in R^{I_\alpha \times I_\beta}$ such that

1.18)
$$\xi(i) = \sum_{\alpha,\beta} \varphi^{\alpha\beta}(i_\alpha, i_\beta).$$

ow if $\{\alpha,\beta\}$ is an adjacent pair $= \{\alpha, \alpha+1\}$ it is easy to check that
$^{\alpha\beta}$ satisfies (1.11) whenever $i,j \in I$ with $i_\mu = j_\mu$. The only non-
rivial case is where $\alpha = \mu$ and then $i_{\{\alpha,\beta\}} = (j:\mu:i)_{\{\alpha,\beta\}}$ and
$_{\{\alpha,\beta\}} = (i:\mu:j)_{\{\alpha,\beta\}}$ because $i_\mu = j_\mu$. This shows that all of the
oefficient vectors are $(,)$ orthogonal to $\mathscr{L}_K$ and so lie in B_K. On
he other hand, if ξ is orthogonal to all of these vectors it
atisfies (1.18) and there cannot exist a pair $\{\alpha,\beta\}$ in the carrier

which is non-adjacent for then there would exist a $\mu \in L$ with
$\alpha < \mu < \beta$ and the coefficient vector of some L_{ij}^{μ} would not be orthogonal to $\varphi^{\alpha\beta}$. Since $\varphi^{\alpha\beta}$ only uses the values of i at the α and
β loci we can choose the i and j to be equal at the μ locus.

QED

So a selection field satisfying only adjacent locus epistasis
preserves the second order recombination ratios of (1.16) and also
certain of the first order recombination ratios of (1.10)--but not
all of them.

The adjacent locus case is also one of the few cases other
than the disjoint bloc generalization of zero epistasis where we can
explicitly compute Λ_K. In fact suppose that $p^{\{\mu,\mu+1\}}$ is an interior
distribution on $I_\mu \times I_{\mu+1}$ and the family $\{p^{\{\mu,\mu+1\}} : \mu = 1,\ldots,\ell-1\}$
is compatible in the sense that $p^{\{\mu,\mu+1\}}$ and $p^{\{\mu-1,\mu\}}$ induce the
same distribution p^μ on I_μ. Then define $p \in \overset{\circ}{\Delta}$ by:

$$(1.19) \qquad p_i = p(i_1,\ldots,i_\ell)$$

$$= \frac{p^{\{1,2\}}(i_1,i_2) \cdot p^{\{2,3\}}(i_2,i_3) \cdot \ldots \cdot p^{\{\ell-1,\ell\}}(i_{\ell-1},i_\ell)}{p^2(i_2) \cdot p^3(i_3) \cdot \ldots \cdot p^{\ell-1}(i_{\ell-1})} .$$

$p_i > 0$ for all $i \in I$ and $E^{\{\mu,\mu+1\}}(p) = p^{\{\mu,\mu+1\}}$ by summing in order
up the loci from 1 to $\mu - 1$ and down from ℓ to $\mu + 2$. At each sum
step only one factor in the numerator is affected and the resulting
sum cancels a factor in the denominator. $p \in \Lambda_K$ by Thm. II.1.6
because $\ell n \, p_i$ is clearly a member of $\mathcal{L}_K$. Notice that in this case
the anamoly about the image of E^K remarked upon after Thm. II.2.7
does not occur.

<u>Disjoint Bloc Model</u> $(K = T_1 \vee \ldots \vee T_{\ell'}$, <u>with</u> $\{T_a: a = 1, \ldots, \ell'\}$ <u>a disjoint family of subsets of</u> L <u>with union</u> L): In this case $\xi \in \mathcal{L}_K$ if it is a sum of terms each depending on the loci in one of the separate blocs.

<u>5 Proposition</u>: For the disjoint bloc case $K = T_1 \vee \ldots \vee T_{\ell'}$, the orthogonal complement B_K is spanned by the coefficient vectors of the family $\{L_{ij}^S: i,j \in I$ and S is a union of some of the T_a's$\}$, i.e. $S \cap T_a = T_a$ or $\emptyset$.

<u>Proof</u>: The special subsets S that we are looking at are precisely those for which S-recombinations do not break up blocs of alleles in K. The L_{ij}^S's are the L_{ij}^λ's corresponding to those maps $\lambda: L \to \{0,1\}$ which are constant on blocs of K. The proof is then an easy analogue of the proof of Lemma II.2.15. The details are left to the reader.

QED

In this case E^K maps Δ onto the product $\Pi_a \Delta_{T_a}$. Furthermore, Λ_K can be explicitly computed. Let p^{T_a} be a distribution on I_{T_a} for $a = 1, \ldots, \ell'$. Since the T_a's are disjoint there are no compatibility conditions necessary and we define $p \in \Delta$:

(1.20) $$P_i = p(i) = p^{T_1}(i_{T_1}) \cdot \ldots \cdot p^{T_{\ell'}}(i_{T_{\ell'}}).$$

p is in Λ_K because $\ell n \ P_i \in \mathcal{L}_K$ (see Thm. II.2.7).

The analogue of Prop. 2 holds with the same proof:

<u>6 Proposition</u>: With respect to the Shahshahani metrics, $E^K: \mathring{\Lambda}_K \to \Pi_a \mathring{\Delta}_{T_a}$ is an isometric diffeomorphism. If m_{ij} is completely symmetric and satisfies K type epistasis then E^K maps $\frac{-}{\nabla}(\frac{1}{2} \ \bar{m})$ on $\mathring{\Lambda}_K$ to the

field which on the factor $\overset{\circ}{\Delta}_{T_a}$ is $\bar{\nabla}(\frac{1}{2} m^{\overline{T_a}})$.

Neutral Loci ($K = S \underset{\neq}{\subset} L$): $\xi \in \mathscr{L}_S$ means that ξ does not depend on the value of the alleles at the loci of the complement $\widetilde{S}$ (cf. II.(2.1)). By Prop. II.2.2, $\mathscr{L}_S$ is the kernel of D_S^j. So by II.(2.5) a spanning set for B_S is given to the coefficient vectors of the family of functions:

(1.21) $\ln p_i/p_j = \ln p_i - \ln p_j$ $i,j \in I$ with $i_S = j_S$.

S type epistasis means that the alleles of the $\widetilde{S}$ loci are neutral, i.e. the fitness numbers do not depend upon them. If a vectorfield on $\overset{\circ}{\Delta}$ has S type epistasis then by (1.21) the ratios $p_{i_S i_{\widetilde{S}}}/p_{i_S j_{\widetilde{S}}}$ remain at some constant value C. Rewrite this as

$$p_{i_S i_{\widetilde{S}}} = C\, p_{i_S j_{\widetilde{S}}}$$

and sum over the i_S loci. We get that $C = p^{\widetilde{S}}(i_{\widetilde{S}})/p^{\widetilde{S}}(j_{\widetilde{S}})$. So for any pair $i_{\widetilde{S}}$, $j_{\widetilde{S}}$ in $I_{\widetilde{S}}$ the ratios

(1.22) $$p^{\widetilde{S}}(i_{\widetilde{S}})/p^{\widetilde{S}}(j_{\widetilde{S}})$$

remain constant under selection satisfying S type epistasis.

One reason for considering these epistasis questions is the hope of reducing the dimension of the problem. The dimension of $\overset{\circ}{\Delta}$ is n-1 which might be quite large. On the other hand $\overset{\circ}{\Delta}_K$, the image of $\overset{\circ}{\Delta}$ under E^K, might be much smaller. If $X = \Sigma\, p_i \xi_i \partial_i$ is a general frequency dependent vectorfield then the image of X at $p \in \overset{\circ}{\Delta}$ under the linear map E^S is computed in Prop. II.2.11. It is:

$$(1.23) \qquad\qquad X^S = \sum P^S_{i_S} \xi^S_{i_S} \partial_{i_S}.$$

X^S is a vector in the tangent space $T_{p^S}(\overset{\circ}{\Delta}_S)$ where $p^S = E^S(p)$ is the distribution induced from p on the subproduct I_S by the projection from I and ξ^S is the conditional expectation $E(\xi|S)$. Just as ξ_i is the fitness of i normalized to mean zero, we call $\xi^S_{i_S}$ the fitness of i_S. It is the average of ξ_i over all of the gametic genotypes which carry the alleles of i_S at the loci of S. In the epistasis zero case with $S = \{\alpha\}$, it is $m(i_\alpha) - \bar{m}$ computed by equation (1.15).

So as p flows in $\overset{\circ}{\Delta}$ along the vectorfield X, p^S flows in $\overset{\circ}{\Delta}_S$ tangent to X^S. However, as (1.15) illustrates, there is an important difference. X^S depends, in general, on the distribution p and not merely on its image p^S. Geneticists refer to this as the effect of "genetic background". It has a rather unpleasant effect: even if we completely understand the genetic background, i.e. we know X^S as a function of p, we can't integrate X^S to get a flow on $\overset{\circ}{\Delta}_S$ because X^S is not a vectorfield on $\overset{\circ}{\Delta}_S$. Here K-type epistasis is helpful in two ways. First, Cor. II.2.12 means that with K-type epistasis we can recover the full vectorfield X from the averaged data $\{X^S: S \in K\}$. Second, if selection alone is acting then in the presence of K-type epistasis each leaf of $\bar{\mathcal{J}}_K$ (e.g. Λ_K, the leaf of maximum entropy) is invariant under the flow. Furthermore, Thm. II. 2.7 implies that E^K maps such a leaf diffeomorphically onto $\overset{\circ}{\Delta}_K$. So in this case we do just flow along $\overset{\circ}{\Delta}_K$ or at least a closely related diffeomorphic image.

If X does not exhibit K-type epistasis we can project orthoogonally (rel. $(\ ,\ )_p$) to $T_p\bar{\mathcal{J}}_K$ to get $X_K = \Sigma\, p(i)\eta(i)\partial_i$. In general, the arguments of Prop. II.2.11 and Cor. II.2.12 show that it is X_K that we can always recover from $\{X^S : S \subset K\}$. Also by the commutative diagram in Prop. II.2.11 (b), η is the $\mathcal{L}_K$ orthogonal projection of ξ. This means that the random variable η_i is the $\mathcal{L}_K$ approximation to ξ_i with least mean square error, i.e. the error $\xi_i - \eta_i$ has minimum variance. Furthermore, the variance of the error is just the square of the length of $X - X_K$ at p with respect to the Shahshahani metric. In the case where $X = \bar{\nabla}(\frac{1}{2} \bar{m})$ then Prop. II.2.14 and the Remark following imply that X_K at p agrees with $\bar{\nabla}_p(\frac{1}{2} \bar{n})$ where n_{ij} is the $\mathcal{L}_{K \times K}$ approximation of m_{ij} with respect to $\bar{p}(\ ,\ )$ $(\bar{p}_{ij} = p_i p_j)$. However, the approximation n_{ij} will vary from point to point. Also, unless p lies in the Wright manifold, the projection from m to n might destroy complete symmetry.

There is an important reason for attempting this sort of reduction in the disjoint bloc case--important beyond the mere convenience of dealing with a small manifold.

As we remarked at the end of Sec. I.1 the vectorfield model is a medium sized model adapted to handling only a small portion of the genome. Our justification for considering it is the hope that we can isolate a medium sized bloc of loci which may interact with one another in determining a component of fitness but which are isolated in their effects from the rest of the genome. In short, we hope that the model we are looking at is a factor of some larger unseen disjoint bloc model. To be precise, suppose that $K = T \vee \tilde{T}$ with $T \subset L$ and what we see are the images of selection, recombination and mutation under the map $E^T \colon \Delta \to \Delta_T$. We call T the set of <u>observed loci</u> and $\tilde{T}$ the set of <u>hidden</u> <u>loci</u>. From this viewpoint it is important to understand not only the dynamics of the vectorfield model on Δ but also how much of the image of these dynamics on Δ_T can be described by the projection of the fields on Δ_T. This is another form of the problem of genetic background.

<u>7 Proposition</u>: Put the Shahshahani metric on $\mathring{\Delta}_T$ as well as $\mathring{\Delta}$ so that $E^T: \mathring{\Delta} \to \mathring{\Delta}_T$ is a Riemannian submersion. Let m be a completely symmetric member of $R^{I \times I}$ exhibiting $T \vee \tilde{T}$ epistasis so that $m = m^T + m^{\tilde{T}}$ where $m^T \in R^{I_T \times I_T}$ and $m^{\tilde{T}} \in R^{I_{\tilde{T}} \times I_{\tilde{T}}}$.

The leaf $\Lambda_{T \vee \tilde{T}}$ consists of those distributions p in $\mathring{\Delta}$ with respect to which the T and $\tilde{T}$ loci are independent, i.e. $d_i^T = 0$ for all i. On $\Lambda_{T \vee \tilde{T}}$ E^T maps the selection field $\bar{\nabla} \frac{1}{2} \bar{m}$ to $\bar{\nabla} \frac{1}{2} \overline{m^T}$ on $\mathring{\Delta}_T$. In general, at p, E^T maps $\bar{\nabla} \frac{1}{2} \bar{m}$ to $\bar{\nabla} \frac{1}{2} \overline{m^T} + \varepsilon(p)$ where the error term is:

$$(1.24) \qquad \varepsilon(p) = \sum \{d_i^T m_{i_{\tilde{T}}}^{\tilde{T}} \partial_{i_T} : i \in I\}.$$

Here $d_i^T = p(i) - p(i_T)p(i_{\tilde{T}})$ and $m_{i_{\tilde{T}}}^{\tilde{T}} = \Sigma\{p(j_{\tilde{T}})m^{\tilde{T}}(i_{\tilde{T}}, j_{\tilde{T}}) : j_{\tilde{T}} \in I_{\tilde{T}}\}$.

If the selection field exhibits T type epistasis, i.e. $m^{\tilde{T}} = 0$ or a constant which can be absorbed in m^T, or equivalently if $\bar{m}$ depends only on the distributions of the alleles in the T loci, then $\bar{\nabla} \frac{1}{2} \bar{m}$ projects under E^T to $\bar{\nabla} \frac{1}{2} \overline{m^T}$ everywhere on $\mathring{\Delta}$.

<u>Proof</u>: The first two paragraphs are excerpts of Thm. II.2.9 and Prop. II.2.11. In particular, (1.24) follows from II.(2.24) and II.(2.25). The S-epistasis case follows from (1.24) directly or from Lemma II. 2.10 applied to the Riemannian submersion E^T. QED

2. Recombination and Entropy.

Throughout this section and the next we will assume that the birth rates b_{ij} and recombination rates r_{ij}^S are nonnegative and completely symmetric. So for $S \subset L$, the recombination field R^S is defined by I. (7.5):

$$(2.1) \qquad R^S = \sum_{i,j} r_{ij}^S b_{ij} d_{ij}^S \partial_i = (1/4) \sum_{i,j} r_{ij}^S b_{ij} d_{ij}^S \bar{\nabla} L_{ij}^S$$

where the functions d_{ij}^S and L_{ij}^S are defined by I. (5.1) and I. (5.2).

The recombination field is $-R$ where (cf. I.(7.2)):

$$(2.2) \qquad R = \sum \{R^S : S \subset L\}.$$

1 Theorem: The vectorfields R and R^S satisfy the following equivalent conditions:

(a) They are tangent to the fibres of the projection $E: \mathring{\Delta} \to \Pi_\alpha \mathring{\Delta}_\alpha$ mapping a distribution to the set of marginal distributions.

(b) They are tangent to the fibre-foliation $\mathfrak{D}$ associated with the zero-epistasis complex $L^{(0)}$.

(c) E maps these vectorfields to zero.

(d) The gene frequencies $p_{i_\alpha}^\alpha$ are integrals of the motion of these vectorfields, i.e. they don't change as p flows according to these vectorfields.

Proof: These conditions are all satisfied by $\bar{\nabla} L_{ij}^S$ and are preserved by linear combination. So the result follows from (2.1).   QEI

The key to our understanding of recombination is the fact that

d_{ij}^S and L_{ij}^S are measuring the same thing in slightly different ways. In particular, they have the same signs. If R^S had L_{ij}^S instead of d_{ij}^S it its formula then R would be a gradient vectorfield because $L_{ij}^S \bar{\nabla} L_{ij}^S = \frac{1}{2} \bar{\nabla}(L_{ij}^S)^2$. We are led to introduce a convenient function.

<u>2 Lemma</u>: Let T be an open interval in R and $f: T \to R$ be a C^∞ function. Define the difference quotient $\Delta_f: T \times T \to R$ by:

$$\Delta_f(s,t) = \begin{cases} \dfrac{f(s) - f(t)}{s - t} & s \neq t \\[2mm] f'(t) & s = t \end{cases}$$

Δ_f is C^∞ and if $s \neq t$ then $\Delta_f(s,t) = f'(\theta)$ for some θ strictly between s and t.

<u>Proof</u>: Changing variables to (t,h) with $s = t + h$ $\Delta_f(t,h) = h^{-1}[f(t + h) - f(t)]$ if $h \neq 0$. So by the integral form of the remainder in Taylor's Theorem

$$\Delta_f(t,h) = \int_0^1 f'(t + hu)\,du$$

and the integral on the right is C^∞ even when $h = 0$.

The last result is the Mean Value Theorem. QED

<u>3 Lemma</u>: Let O be the first quadrant of the plant R^2, i.e. $O = \{(s,t): s,t \geq 0\}$ and let $\overset{\circ}{O}$ be the interior where the inequalities are strict. On O we define

$$
Q(s,t) = \begin{cases} \dfrac{s-t}{\ln s - \ln t} & s \neq t \text{ and } s \cdot t > 0 \\[2mm] t & s = t > 0 \\[2mm] 0 & s \cdot t = 0 \end{cases}
$$

Q satisfies the following properties.

(a) Q is continuous on O and is infinitely differentiable on $\overset{\circ}{O}$.

(b) Q is non-negative with $\overset{\circ}{O} = \{Q > 0\}$.

(c) $Q(s,t)$ is between s and t, strictly between if $s \neq t$ and $s \cdot t > 0$.

(d) Q is homogeneous of degree 1, i.e. $Q(\lambda s, \lambda t) = \lambda Q(s,t)$ for $\lambda \geq 0$.

(e) Q is symmetric, i.e. $Q(s,t) = Q(t,s)$.

(f) $\dfrac{\partial Q}{\partial s}(s,t) = \dfrac{\partial Q}{\partial t}(t,s)$ are positive and homogeneous of degree 0 on $\overset{\circ}{O}$.

Proof: Q is the reciprocal of the difference quotient for the logarithm $\ln$ defined on the positive reals. Since the derivative of $\ln t$ is $1/t$ (a), (b) and (c) on $\overset{\circ}{O}$ follow from Lemma 1. (d) and (e) are obvious.

If s approaches 0 while t remains bounded away from 0 and ∞, $Q(s,t)$ approaches 0 like $s/\ln s$. By symmetry, Q goes to 0 if t approaches 0 alone. If both s and t approach 0 then they take $Q(s,t)$, which lies between them, along.

Finally, for (f):

$$
\frac{\partial Q}{\partial s} = \frac{(t/s) - \ln(t/s) - 1}{(\ln(t/s))^2} \qquad (s \neq t).
$$

This is positive because $r - 1 > \ell n\, r$ for all positive $r \neq 1$. When $s = t$, two applications of L'Hopital's Rule imply that $\frac{\partial Q}{\partial s} = \frac{1}{2}$. QED

4 Proposition: On Δ define the function $Q^S_{ij} = Q(p_i p_j, p_{\overline{i}} p_{\overline{j}})$, i.e. $= Q(p(i)p(j), p(i_S j_{\widetilde{S}})p(j_S i_{\widetilde{S}}))$. Q^S_{ij} is continuous on Δ and C^∞ on $\mathring{\Delta}$. Furthermore Q^S_{ij} is positive on $\mathring{\Delta}$. Since $Q^S_{ij} = d^S_{ij}/L^S_{ij}$:

$$(2.3) \qquad R^S = (1/8) \sum_{i,j} r^S_{ij} b_{ij} Q^S_{ij} \overline{\nabla}(L^S_{ij})^2.$$

Proof: This is clear from Lemma 3, (1.2) and the fact that $\overline{\nabla}(L^S_{ij})^2 = 2\, L^S_{ij} \overline{\nabla} L^S_{ij}$. QED

Remark: Since d^S_{ij} and L^S_{ij} are antisymmetric in the variables (i_S, j_S) and in $(i_{\widetilde{S}}, j_{\widetilde{S}})$, the quotient Q^S_{ij} and the square $(L^S_{ij})^2$ are symmetric in these variables. The hypothesis of complete symmetry says that b_{ij} and r^S_{ij} are also symmetric in these variables.

Recall the definition of entropy $H: \Delta \to [0,\infty)$, continuous on Δ and C^∞ on $\mathring{\Delta}$:

$$(2.4) \qquad H(p) = -\sum p_i\, \ell n\, p_i = -\frac{1}{2}\sum p_i p_j\, \ell n\, p_i p_j.$$

For each marginal distribution p^α on Δ_α we define $H^\alpha: \Delta \to [0,\infty)$

$$(2.5) \qquad H^\alpha(p) = -\sum p_i\, \ell n\, p^\alpha(i_\alpha) = -\sum p(i_\alpha)\, \ell n\, p^\alpha(i_\alpha)$$

$$= -\frac{1}{2}\sum p_i p_j\, \ell n\, p^\alpha(i_\alpha)p^\alpha(j_\alpha).$$

We define the product distribution function $\pi: \Delta \to \Delta$ mapping $\mathring{\Delta}$ into $\mathring{\Delta}$ by

$$(2.6) \qquad \pi(p)_i = \pi_i = \prod_{\alpha=1}^{\ell} p^\alpha(i_\alpha).$$

So we can define the normalized entropy $\hat{H}$ (cf. II.(1.11)):

$$(1.7) \qquad \hat{H}(p) = H(p) - \sum_{\alpha=1}^{\ell} H^{\alpha}(p) = H(p) - H(\pi)$$

$$= -\frac{1}{2} \sum p_i p_j \, \ln(p_i p_j / \pi_i \pi_j).$$

$\hat{H}$ is nonpositive on $\overset{\circ}{\Delta}$ and vanishes on $\overset{\circ}{\Lambda}$ by Lemma II.1.7. Note that the H^{α}'s depend only on the marginal distributions and so these functions factor through $E: \Delta \to \Pi_{\alpha} \Delta_{\alpha}$.

5 Theorem: The following equation holds on $\overset{\circ}{\Delta}$:

$$(2.8) \qquad (-R, \bar{\nabla}_p H)_p = (-R, \bar{\nabla}_p \hat{H})_p = (1/4) \sum_{i,j,S} r^S_{ij} b_{ij} Q^S_{ij} (L^S_{ij})^2.$$

    (a) The sum is nonnegative and vanishes exactly when R vanishes.

    (b) R vanishes on the Wright manifold, $\overset{\circ}{\Lambda}$.

    (c) $\bar{\nabla}_p H$ vanishes exactly at the point $p \in \overset{\circ}{\Lambda}$ where $p_i = 1/n$.

    (d) $\bar{\nabla}\hat{H}$ vanishes exactly on the Wright manifold $\overset{\circ}{\Lambda}$.

In particular, if R vanishes only on $\overset{\circ}{\Lambda}$, then $\hat{H}$ is a Lyapunov function for $-R$.

Proof: (2.8) follows from (2.3) by rewriting $\bar{\nabla}L^2 = 2L\bar{\nabla}L$ and applying the equation $(\bar{\nabla}_p H, \bar{\nabla}_p L)_p = L(p)$ (cf. I.(4.13)). Note that $\bar{\nabla}H$ and $\bar{\nabla}\hat{H}$ differ by the sum $\Sigma_{\alpha} \bar{\nabla}H^{\alpha}$ which is parallel to the transverse foliation $\mathcal{F}$ because the H^{α}'s factor through E. So this sum is perpendicular to the $\bar{\nabla}L^S_{ij}$'s, and $(\bar{\nabla}_p H, \bar{\nabla}_p L)_p = (\bar{\nabla}_p \hat{H}, \bar{\nabla}_p L)_p$.

Since Q^S_{ij} is positive on $\overset{\circ}{\Delta}$ the sum in (2.8) is nonnegative and vanishes only when for all S, i and j $r^S_{ij} b_{ij}$ or L^S_{ij} vanish.

n that case R vanishes. R vanishes on $\overset{\circ}{\Lambda}$ because all of the

S_{ij}'s do. This proves (a) and (b).

For (c), $\bar{\nabla}_p H = -\Sigma\, p_i(\ln p_i + H)\partial_i$ and so $\bar{\nabla}_p H$ vanishes in $\overset{\circ}{\Delta}$

only when all the $\ln p_i$'s are equal.

(d) follows from Lemma II.1.7.   QED

The question of whether R vanishes only on $\overset{\circ}{\Lambda}$ is resolved,

in principle, by the following:

Proposition: R vanishes only on $\overset{\circ}{\Lambda}$ iff the coefficient vectors

of the set of functions $\{L^S_{ij} : r^S_{ij}, b_{ij} > 0\}$ span the orthogonal comple-

ment $B^{(0)}$ of the zero epistasis subspace $\mathcal{L}^{(0)}$. This occurs iff for

every $p \in \overset{\circ}{\Delta}$ the set of vectors $\{\bar{\nabla}_p L^S_{ij} : r^S_{ij} b_{ij} > 0\}$ spans the tangent

space of the fibre foliation $\mathfrak{D}$.

For example, if $r^S_{ij} > 0$ for all $S = S_\mu = \{\alpha \leq \mu\}$ $\mu = 1, \ldots, \ell$

and there exists $j \in I$ such that $b_{ij} > 0$ for all i, then R vanishes

only on $\overset{\circ}{\Lambda}$.

Proof: This result is based on the constructions for Thm. II.1.1, and

the special case of zero-epistasis from the previous Sec. Recall

that for $b \in R^I$ we defined $L^b : \overset{\circ}{\Delta} \to R$ by $L^b(p) = \Sigma\, b_i \ln p_i$. A linear

relation among the b's is equivalent to a linear relation among the

corresponding L^b's or again among the corresponding $\bar{\nabla}_p L^b$'s at every

point p of $\overset{\circ}{\Delta}$. The equivalence of the two different conditions in

the first paragraph follows from the fact that the map $b \to \bar{\nabla}_p L^b$ is an

isomorphism of B onto the tangent space $T_p \mathfrak{D}$ (cf. II.(1.5)). Now

let $\tilde{B}$ be the subspace of R^I spanned by the coefficient vectors of

$\{L^S_{ij} : r^S_{ij}, b_{ij} > 0\}$. Since $B^{(0)}$ is spanned by coefficient vectors of all of the L_{ij}^S's, $\tilde{B} \subset B^{(0)}$. Notice that R vanishes at p iff $L_{ij}^S(p) = 0$ whenever $r_{ij}^S \, b_{ij} > 0$.

Suppose $\tilde{B} = B^{(0)}$ and R vanishes at $p \in \mathring{\Delta}$. All of the L_{ij}^S's are linear combinations of those with $r_{ij}^S \, b_{ij} > 0$ because $\tilde{B} = B^{(0)}$. So $L_{ij}^S(p) = 0$ for all i,j and S. Hence, $p \in \mathring{\Lambda}$.

On the other hand if $\tilde{B}$ is a proper subspace of B let $\tilde{A}$ be the orthogonal complement of $\tilde{B}$. $\mathcal{L}^{(0)}$ is a proper subspace of $\tilde{A}$. In the notation of Thm. II.1.1 R vanishes at $p \in \mathring{\Delta}$ iff $L^{\tilde{B}}(p) = 0$ and by that theorem $(L^{\tilde{B}})^{-1}(0)$ is a submanifold of $\mathring{\Delta}$ containing $\mathring{\Lambda}$ but of higher dimension. In this case the maps $E^a(p) = \Sigma \, a_i p_i$ for a in $\tilde{A} - \mathcal{L}^{(0)}$ give integrals of motion for the recombination flow, additional to the individual gene frequencies--cf. Thm. 1 (d)--which come from $\mathcal{L}^{(0)}$.

In discussing the zero-epistasis example, we saw that the coefficient vectors of L_{ij}^μ, for $i \in I$, $\mu \in L$ with a fixed j span $B^{(0)}$. This proves the example. QED

There is little harm in assuming that $r_{ij}^S > 0$ for all i,j at least with $S = S_\mu$, $\mu = 1,\ldots,\ell$. When b_{ij} is zero rather than positive we say that the ij zygote is _sterile_. A simple example where R vanishes on more than $\mathring{\Lambda}$ is the case of a _dominant lethal gamete-type_, $\tilde{I}$, meaning $b_{\tilde{I}j} = 0$ for all $j \in I$. If i_α is a dominant lethal gene then every gamete i with this allele at the α locus is a dominant lethal gamete-type. To see that R vanishes on more than $\mathring{\Lambda}$ we will show $\mathcal{L}^{(0)}$ is a proper subspace of $\tilde{A}$, in the notation of the above proof. In fact if $a_k = \delta_{\tilde{I}k}$ (the Kronecker delta) then E^a maps R to 0. This is because, by complete symmetry the coefficient of $\partial_{\tilde{I}}$ in R^S is zero if $b_{ij} > 0$. Thus, projecting the selection-plus-

recombination vectorfield by E^a eliminates R, i.e. the frequency $p_{\tilde{i}}$
is affected only by selection:

(2.9)
$$\frac{dp_{\tilde{i}}}{dt} = p_{\tilde{i}}(m_{\tilde{i}} - \bar{m}).$$

Now to a biologist what happens is clear. A dominant lethal is
simply eliminated from the population by selection. This suggests
the conjecture that when there is sufficient sterility that R vanishes on more than $\dot{\Lambda}$ the population is driven out of the interior
of $\dot{\Delta}$ and all orbits of the selection-plus-recombination field
approach the boundary: $\Delta - \dot{\Delta}$. But for a moment this is mathematically
puzzling. The relation between sterility and selection is: $b_{ij} = 0$
implies $m_{ij} < 0$. This is because $m_{ij} = b_{ij} - d_{ij}$ and d_{ij} is assumed
> 0 for all i,j (no immortality). But the sign of m_{ij} is irrelevant
to the selection field on $\dot{\Delta}$. Addition of any constant to all of the
m_{ij}'s doesn't affect selection on genotype frequencies. So in theory
a dominant lethal could be selected for. The patient biologist then
points out that the dominant lethal increases in frequency under
selection only if everything else is being eliminated even faster.
Indeed from (2.9), since $m_{\tilde{i}} = \Sigma\, p_j m_{\tilde{i}j} < 0$, one of the following must
be true: $\frac{dp_{\tilde{i}}}{dt} < 0$, $\bar{m} < 0$ or $p_{\tilde{i}} = 0$. This leads to:

The Sterility Conjecture: If $\{\bar{\nabla}L_{ij}^S : b_{ij} > 0\}$ does not span the tangent space $T\mathfrak{D}$, then from every initial position the flow of the selection-plus-recombination field approaches the boundary $\Delta - \dot{\Delta}$ or the
population size approaches 0 (extinction) where the population
size $|x|$ satisfies the equation:

(2.10)
$$\frac{d|x|}{dt} = |x|\bar{m}.$$

For a dominant lethal, $\tilde{\imath}$, this is proved by looking not at (2.9) but at the equation for the absolute number (cf. I.(1.1)):

$$(2.11) \qquad \frac{dx_{\tilde{\imath}}}{dt} = x_{\tilde{\imath}} m_{\tilde{\imath}} \ .$$

Since $m_{\tilde{\imath}} < 0$, $x_{\tilde{\imath}}$ approaches 0. Hence, either the frequency $p_{\tilde{\imath}} = x_{\tilde{\imath}}/|x|$ approaches 0 or $|x|$ does.

From now on we will simply assume that R vanishes only on $\dot{\Lambda}$, and so normalized entropy, $\hat{H}$, is a Lyapunov function for $-R$.

Since we know that $\bar{m}$ increases under selection and H increases under recombination, it is of interest to consider the opposite pairing and see how $\bar{m}$ behaves under recombination and H under selection.

If there is no epistasis then the gradient of $\bar{m}$ is parallel to the transverse foliation $\bar{\mathcal{F}}$ and so is perpendicular to the $\bar{\nabla}L_{ij}^{S}$'s and to $-R$. The extent to which there is epistasis is measured by the functions $e_{ij}^{S} : \Delta \to R$ defined as follows:

$$(2.12) \qquad e_{ij}^{S} = (\bar{\nabla}_{p} L_{ij}^{S}, \bar{\nabla}_{p} \tfrac{1}{2} \bar{m})_{p}$$

$$= m_{i} - m_{\bar{\imath}} - m_{\bar{\jmath}} + m_{j} \qquad (\bar{\imath} = i_{S} j_{\tilde{S}}, \bar{\jmath} = j_{S} i_{\tilde{S}}).$$

These are linear functions, as the gamete frequencies are. In fact we can define the numbers:

$$(2.13) \qquad e_{ij,k}^{S} = (\bar{\nabla}_{p} L_{ij}^{S}, \bar{\nabla}_{p} m_{k})_{p} = m_{ik} - m_{\bar{\imath}k} - m_{\bar{\jmath}k} + m_{jk}$$

and then we clearly have

2.14)
$$e_{ij}^S(p) = \sum_K p_k e_{ij,k}^S .$$

The cumulative effect of epistasis, $e: \Delta \to R$ which measures the effect of recombination on fitness is given by:

2.15)
$$e = (R, \bar{\nabla}_p \bar{m})_p$$

$$= (1/2) \sum_{i,j,S} r_{ij}^S b_{ij} d_{ij}^S e_{ij}^S$$

$$= (1/2) \sum_{i,j,S} r_{ij}^S b_{ij} Q_{ij}^S L_{ij}^S e_{ij}^S .$$

So the selection-plus-recombination field acts on mean fitness by the formula:

2.16)
$$(\bar{\nabla}_p \frac{1}{2} \bar{m} - R, \bar{\nabla}_p \bar{m})_p = V_A - e.$$

Here the first term on the left is the additive variance
$V_A = 2\Sigma_i p_i (m_i - \bar{m})^2 = (\bar{\nabla}_p \frac{1}{2} \bar{m}, \bar{\nabla}_p \bar{m})$ (see I.(6.3) and I.(6.9)).

At $p \in \overset{\circ}{\Delta}$ we can consider $i \to \ell n \, p_i$ as a random variable and so define the covariance of $\ell n \, p_i$ with fitness:

2.17)
$$\text{Cov}(\ell n \, p, m) \equiv \sum_i p_i (\ell n \, p_i)(m_i - \bar{m}) = \sum_i p_i (\ell n \, p_i) m_i + H\bar{m}$$

$$= [\frac{1}{2} \sum_{i,j} p_i p_j (\ell n \, p_i p_j) m_{ij}] + H\bar{m}.$$

Where H is defined by (1.5) and $\bar{m} = \Sigma_i p_i m_i = \Sigma p_i p_j m_{ij}$. Note that because $\ell n(p_i p_j) = \ell n \, p_i + \ell n \, p_j$ the covariance of $\ell n(p_i p_j)$ with m_{ij} (relative to $p_i p_j$) is just twice the covariance of $\ell n \, p_i$ with m_i (relative to p_i).

Direct substitution using the formula for $\bar{\nabla}H$ yields:

(2.18) $\qquad\qquad\qquad (\bar{\nabla}\,\frac{1}{2}\,\bar{m},\bar{\nabla}H) = -\text{Cov}(\ell n\ p, m).$

7 Proposition: Let p be any initial position in $\overset{\bullet}{\Delta}$. Let $\{p_t : t \geq 0\} \subset \overset{\bullet}{\Delta}$ be the positive orbit of p under the selection-plus-recombination field, $\bar{\nabla}\,\frac{1}{2}\,\bar{m} - R$.

$$\lim \sup_{t\to\infty} \text{Cov}(\ell n\ p, m)(p_t) \geq 0$$

(2.19)

$$\lim \sup_{t\to\infty} e(p_t) \geq 0.$$

In particular, if p_t approaches a limit p_∞, an equilibrium for the vectorfield, then we can replace lim sup by lim in (2.19). If $p_\infty \in \overset{\bullet}{\Delta}$ then these limits are both positive (or both zero) if $p_\infty \notin \Lambda$ (resp. if $p_\infty \in \overset{\bullet}{\Lambda}$), provided R vanishes only on $\overset{\bullet}{\Lambda}$.

Proof: For $f = \bar{m}$ or H, f is bounded and smooth on $\overset{\bullet}{\Delta}$ and so the derivative along the path, $\frac{df}{dt} = (\bar{\nabla}_{p_t}\,\frac{1}{2}\,\bar{m} - R, \bar{\nabla}_{p_t} f)_{p_t}$ can't be bounded above zero, i.e. it cannot happen that $\frac{df}{dt} > \epsilon$ for all $t > t_0$ and any $\epsilon > 0$. This means Lim inf $\frac{df}{dt} \leq 0$. For $f = \frac{1}{2}\,\bar{m}$, for example, $V_A \geq 0$ implies:

$$- \lim \sup\ e \leq \lim \inf \frac{df}{dt} \leq 0.$$

This proves the second inequality in (2.19). The first is similar using positivity of $(-R, \bar{\nabla}_p H)_p$.

If p_t approaches p_∞ then by continuity the limits in (2.19) exist and equal $\text{Cov}(\ell n\ p, m)(p_\infty)$ and $e(p_\infty)$. At an internal equilibrium $\bar{\nabla}\,\frac{1}{2}\,\bar{m} - R$ vanishes and so the left sides are zero iff V_A and $(-R, \bar{\nabla}_p H)_p$ vanish which they do exactly when the equilibrium is in $\overset{\bullet}{\Lambda}$.

$\qquad\qquad\qquad\qquad\qquad\qquad\qquad\qquad\qquad\qquad\qquad\qquad\qquad$ QED

Remark: Here we have used the fact that $\bar{\nabla}_p \frac{1}{2} \bar{m} - R$ is parallel to or pointing inward on each face of Δ and so by compactness arguments the positive orbit p_t is defined for all positive t and lies in $\overset{\bullet}{\Delta}$ if p does.

That $Cov(\ln p, m)$ should tend to be positive is intuitively appealing. A positive corellation between $\ln p_i$ and m_i means that the more fit genotypes are relatively more frequent and one would expect this effect to be intensified by selection. This argument is misleading. Under selection alone every orbit tends to an equilibrium at which all of the genotypes which occur have the same fitness, i.e. $V_A = 0$. So fitness m_i tends to become uncorrelated with anything. To suggest that recombination is improving on selection by possibly allowing $Cov(\ln p, m)$ to remain positive is probably a misinterpretation of the results.

That e should tend to be positive is a weak generalization of Felsenstein's results in [11] suggesting that e_{ij}^S and d_{ij}^S tend eventually to have the same sign. This interpretation is correct in the two-locus-two-allele model, where the sum in e has essentially only one term. In general, e is a large sum and we can't say that all of the terms are positive.

3. Recombination and Epistasis.

In this section we examine the conditions under which the recombination field is tangent to the maximum entropy leaf Λ_K of the transverse foliation $\bar{\mathcal{F}}_K$ associated with K type epistasis. We will also see why this tangency usually does not hold.

The results are exhibited most clearly in the case where the birth rates and recombination rates are genotype independent.

<u>1 Proposition</u>: For $S \subset L$ $i \in I$ define $d_i^S: \Delta \to R$ by:

(3.1)
$$d_i^S = \sum_j d_{ij}^S = p(i) - p(i_S)p(i_{\widetilde{S}}).$$

Assume $b_{ij} = b$ and $r_{ij}^S = r^S$. Then R^S is given by:

(3.2)
$$R^S = r^S b \sum_i d_i^S \partial_i.$$

If K is a complex of subsets of L, then $p \in \Lambda_K$ iff $\ell n \ p_i \in \mathcal{L}_K$ as a function of i. R^S is tangent to the transverse foliation $\bar{\mathcal{J}}_K$ at p iff $p(i_S)p(i_{\widetilde{S}})/p(i) \in \mathcal{L}_K$ as a function of i.

<u>Proof</u>: (3.2) is clear from (2.1). The criterion for $p \in \Lambda_K$ comes from Thm. II.1.6. By Addendum II.1.3 R^S is tangent to $\bar{\mathcal{J}}_K$ at p iff, as a function of i, $d_i^S/p(i)$ lies in $\mathcal{L}_K$. Since $\mathcal{L}_K$ contains the constant functions, this is true iff $p(i_S)p(i_{\widetilde{S}})/p(i)$ lies in $\mathcal{L}_K$.   QED

Now $\ell n \ p_i \in \mathcal{L}_K$ means that p_i is a product of functions in $\mathcal{L}_S$ for $S \in K$. This will sometimes imply that $p(i_S)p(i_{\widetilde{S}})/p(i)$ is a similar product, but only rarely that it is a sum of functions in $\mathcal{L}_S$ for $S \in K$. For example, in the disjoint bloc case, $K = T_1 \vee \ldots \vee T_{\ell'}$, define $S_a = S \cap T_a$ and $\widetilde{S}_a = (\widetilde{S})_a = \widetilde{S} \cap T_a$ for $a = 1,\ldots,\ell'$. If $p \in \Lambda_K$ then independence implies $p(i_S) = \Pi_a p(i_{S_a})$ and so

(3.3) $p(i_S)p(i_{\widetilde{S}})/p(i) = \Pi_a p(i_{S_a})p(i_{\widetilde{S}_a})/p(i_{T_a})$ $(p \in \Lambda_K)$.

Now the log of this function lies in $\mathcal{L}_K$ but the function itself usuall does not.

<u>2 Corollary</u>: Assume $b_{ij} = b$, $r^S_{ij} = r^S$ and $K = T_1 \vee \ldots \vee T_{\ell'}$ is a disjoint bloc model. If $S \supset T_a$ or $\tilde{S} \supset T_a$ for all but at most one of $a = 1, \ldots, \ell'$ then R^S is tangent to Λ_K at all point of Λ_K.

<u>Proof</u>: In this case $S_a = T_a$ or $\tilde{S}_a = T_a$ for all a but say a_0. So in the product on the right of (3.3) all of the factors equal 1 except for the a_0 factor which depends only on i_T with $T = T_{a_0}$. QED

<u>3 Corollary</u>: Assume $b_{ij} = b$, $r^S_{ij} = r^S$ and $K = \{1,2\} \vee \{2,3\} \vee \ldots \ldots \vee \{\ell-1, \ell\}$ is the adjacent locus interaction mode. If $S = S_\mu = \{\alpha \in L: \alpha \leq \mu\}$ for some $\mu \in L$ then R^S is tangent to Λ_K at all points of Λ_K.

<u>Proof</u>: By computing with (1.19) it is not hard to show that for $S = S_\mu$ and $p \in \Lambda_K$:

$$(p(i_S)p(i_{\tilde{S}}))/p(i) = (p^\mu(i_\mu)p^{\mu+1}(i_{\mu+1}))/p^{\{\mu,\mu+1\}}(i_\mu i_{\mu+1}).$$

The function on the right depends only on the pair of loci $\{\mu, \mu+1\} \in K$. So it is a function in $\mathscr{L}_K$. QED

There is another very special case of the disjoint bloc model where Λ_K tangency holds even for the more general recombination fields of (2.1).

<u>4 Lemma:</u> Let $S.T \subset L$. If $p \in \Lambda_{T \vee \tilde{T}}$, then

$$(3.4) \quad d^S_{ij} = p(i_{\tilde{T}})p(j_{\tilde{T}})d^{S \cap T}_{i_T j_T} + p(\bar{i}_T)p(\bar{j}_T)d^{S \cap \tilde{T}}_{i_{\tilde{T}} j_{\tilde{T}}}$$

$$= p(i_{\tilde{T}})p(j_{\tilde{T}})d^{S \cap T}_{i_T j_T} + p(i_T)p(j_T)d^{S \cap \tilde{T}}_{i_{\tilde{T}} j_{\tilde{T}}} - d^{S \cap T}_{i_T j_T} d^{S \cap \tilde{T}}_{i_{\tilde{T}} j_{\tilde{T}}}$$

where $\bar{i} = i_S j_{\tilde{S}}$ and $\bar{j} = j_S i_{\tilde{S}}$ and $d^{S \cap T}_{i_T j_T}$ is the analogue of d^S_{ij} on Δ_T.

In the last sum the first and third terms vanish if

$p \in \Lambda_{T}{}^{(0)}{}_{\vee \tilde{T}}$ or if $S \subset \tilde{T}$. The second and third terms vanish if

$p \in \Lambda_{T \vee \tilde{T}}{}^{(0)}$ or if $S \subset T$. $T^{(0)}$ is the complex of singleton subsets

of T.

<u>Proof</u>: By (1.20), if $p \in \Lambda_{T \vee \tilde{T}}$, then

$$d^S_{ij} = p(i_{\tilde{T}}) p(j_{\tilde{T}}) p(i_T) p(j_T) - p(\bar{i}_{\tilde{T}}) p(\bar{j}_{\tilde{T}}) p(\bar{i}_T) p(\bar{j}_T).$$

separate the two terms. Then add and subtract $p(i_{\tilde{T}}) p(j_{\tilde{T}}) p(\bar{i}_T) p(\bar{j}_T)$

to get (3.4). If $p \in \Lambda_{T}{}^{(0)}{}_{\vee \tilde{T}}$ then the loci of T are all independent

and so $d^{S \cap T}_{i_T j_T} = 0$. If $S \subset \tilde{T}$ then $S \cap \tilde{T} = S$ and $S \cap T = \emptyset$. $d^{\emptyset}$ is

always 0. Similarly, for the complementary cases.   QED

Now apply Addendum II.1.3 just as in the proof of Proposition

1. If $r^S_{ij} = r^S$ then R^S is tangent to Λ_K at p iff $\Sigma_j b_{ij} d^S_{ij}/P_i$ lies

in $\mathcal{L}_K$ as a function of i. Now suppose that $p \in \Lambda_{T \vee \tilde{T}}$ and

$b = b^T \in R^{I_T \times I_T}$. Then we can sum on j by summing first on $j_{\tilde{T}}$ and

then on j_T to get:

$$(3.5) \quad \sum_j b^T_{i_T j_T} d^S_{ij} = p(i_{\tilde{T}}) \sum_{j_T} b^T_{i_T j_T} + d^{S \cap \tilde{T}}_{i_{\tilde{T}}} [p(i_T) b^T_{i_T} - \sum_{j_T} b^T_{i_T j_T} d^{S \cap T}_{i_T j_T}],$$

where

$$b^T_{i_T} = \sum_j P_j b^T_{i_T j_T} = \sum_{j_T} P_{j_T} b^T_{i_T j_T}$$

and

$$d^{S \cap \tilde{T}}_{i_{\tilde{T}}} = \sum_{j_{\tilde{T}}} d^{S \cap \tilde{T}}_{i_{\tilde{T}} j_{\tilde{T}}}$$

(following (3.1)).

If we divide by $p(i) = p(i_T)p(i_{\tilde{T}})$ the first term then depends only on i_T. The second is the product of an i_T function and an $i_{\tilde{T}}$ function, which causes the problem. However, if $p \in \Lambda_{T \vee \tilde{T}}(0)$ then $d_{i_{\tilde{T}}}^{S \cap \tilde{T}} = 0$ and so the second term doesn't occur. By the same argument with b_{ij} replaced by $r_{ij}^S b_{ij}$ we prove:

<u>5 Proposition</u>: Assume that r_{ij}^S and b_{ij} are completely symmetric members of $R^{I_T \times I_T}$, i.e. the $\tilde{T}$ loci are neutral with respect to birth rates and recombination rates. If $K = T \vee \tilde{T}^{(0)}$ then the recombination fields R^S are tangent to Λ_K at all points of Λ_K for all $S \subset L$.

In order to apply these results, we compute the image of the recombination fields under the projection $E^T: \Delta \to \Delta_T$.

<u>6 Proposition</u>:

$$(3.6) \qquad E^T(\bar{\nabla}L_{ij}^{S'}) = \bar{\nabla}L_{i_T j_T}^S \qquad (S = S' \cap T)$$

where the gradient on the right is taken with respect to the Shahshahani metric on $\overset{\bullet}{\Delta}_T$.

Assume $r_{ij}^{S'} = r_{i_T j_T}^{S'}$ for all i and j. Then

$$(3.7) \quad E^T(R^{S'}) = \sum_{i_T, j_T} r_{i_T j_T}^{S'} (E(b|i_T, j_T)p_{i_T}p_{j_T} - E(b|\bar{i}_T, \bar{j}_T)p_{\bar{i}_T}p_{\bar{j}_T})\partial_{\bar{i}_T}$$

$$= (1/4) \sum_{i_T, j_T} r_{i_T j_T}^{S'} (E(b|i_T, j_T)p_{i_T}p_{j_T}$$

$$- E(b|\bar{i}_T, \bar{j}_T)p_{\bar{i}_T}p_{\bar{j}_T})\bar{\nabla}L_{i_T j_T}^S.$$

Where $\mathbf{E}(b|i_T,j_T)$ is the conditional expectation of b_{ij} assuming i_T ar j_T are known, with distribution $p_i p_j$ on $I \times I$. $\bar{i}_T = i_{S' \cap T} j_{\tilde{S}' \cap T}$ and $S = S' \cap T$.

So $E^T(\Sigma\{R^{S'} : S' \cap T = S\})$ is given by the same sums with $r^{S'}_{i_T j_T}$ replaced by $r^{S,T}_{i_T j_T} = \Sigma\{r^{S'}_{i_T j_T} : S' \cap T = S\}$.

<u>Proof:</u> Since $E^T(\partial_i) = \partial_{i_T}$ and $(\bar{i})_T = \bar{i}_T$, i.e. $(i_S j_{\tilde{S}})_T = i_S j_{T-S}$, (3.6) is clear. From this (3.7) follows because:

$$\sum \{b_{k\ell} p_k p_\ell : k_T = i_T \text{ and } \ell_T = j_T\} = \mathbf{E}(b|i_T,j_T) p_{i_T} p_{j_T}$$

by definition of conditional expectation. QEI

<u>Remarks:</u> (a) It follows from (3.6) and Lemma II.1.10 that the hori-zontal projection of $\bar{\nabla} L^{S'}_{ij}$, i.e. the projection of $\bar{\nabla} L^{S'}_{ij}$ perpendicular to the fibres of E^T, is given by $\bar{\nabla}(L^S_{i_T j_T} \cdot E^T)$. For by Prop. II.1.11(l E^T is a Riemannian submersion.

(b) The form of (3.7) is similar to I.(7.1) or I.(7.4) but noi necessarily to I.(7.5) even if b_{ij} was initially completely symmetric If b depends only on i_T and j_T, i.e. $b = b^T \in R^{I_T \times I_T}$ then $\mathbf{E}(b|i_T,j_T) = b^T_{i_T j_T}$. Also, if the T and $\tilde{T}$ loci are independent, i.e. $p \in \Lambda_{T \vee \tilde{T}}$, Prop. II.1.14(a) implies that $\mathbf{E}(b|T)$ is completely symmetric if b is. But away from $\Lambda_{T \vee \tilde{T}}$ we may lose complete symmeti by projecting and so have <u>observed</u> <u>position</u> <u>effects</u> of the projectiec field even if there were no position effects in the original.

<u>7 Corollary:</u> Let $K = T_1 \vee \ldots \vee T_\ell$, be a disjoint bloc complex. For $S \subset L$, define $S_a = S \cap T_a$. Assume $r^S_{ij} = r^S$ for all S and $b_{ij} = b$. Define R^S_a to be the recombination field for S_a (as in (3.2)) but witl

r^{S_a} replaced by $r^{S_a, T_a} = \Sigma\{r^{S'} : S' \cap T_a = S_a\}$. Then

(a) $E^K(R^S) = E^K(\Sigma_{a=1}^{\ell'} R_a^S)$.

(b) $\Sigma_{a=1}^{\ell'} R_a^S$ is tangent to Λ_K at all points of Λ_K.

(c) At $p \in \Lambda_K$, $\Sigma_{a=1}^{\ell'} R_a^S$ is the $(\ ,\ )_p$ orthogonal projection of R^S on $T_p\Lambda_K$.

<u>Proof:</u> E^K is the product of the maps E^T with $T = T_a$. (3.7) makes it clear that $E^{T_b}(R_a^S) = 0$ if $a \neq b$ and $E^{T_a}(R_a^S) = E^{T_a}(R^S)$. (a) follows. (b) follows from Cor. 2. (c) follows from (a) and (b) and Prop. II.2.11(b). $\hspace{2cm}$ QED

<u>Remarks:</u> (a) This result illustrates again that tangency problems arise from recombination occurring in more than one bloc at once.

(b) Note that recombination among blocs is invisible with respect to E^K, i.e. if $S_a = T_a$ or $\emptyset$ for all a, then $E^K(R^S) = 0$ and $R^S|_{\Lambda_K} = 0$, assuming as above that the birth rates are constant.

As was remarked at the end of Sec. 1, it is best to regard the vectorfield model as part of a larger disjoint bloc model. It then becomes important to study the relation between the large model and its projection to Δ_T. Recall that we call the loci in T the observed loci and the remaining loci, those in $\tilde{T}$, the hidden loci. By the <u>observed recombination</u> or <u>selection field</u> we will mean the image of the recombination or selection field under the projection $E^T: \Delta \to \Delta_T$.

<u>Birth and Recombination Rates Independent of Hidden Loci;</u>

<u>Hidden Loci Contribute Additively to Death Rates:</u> This means that r_{ij}^S and b_{ij} are completely symmetric members of $R^{I_T \times I_T}$ and d_{ij} shows

$T \vee \tilde{T}^{(0)}$ type epistasis. So m_{ij} shows $T \vee \tilde{T}^{(0)}$ type epistasis and the selection field is tangent to $\Lambda_{T\vee\tilde{T}(0)}$. By Corollary 5 the recombination fields are also tangent to $\Lambda_{T\vee\tilde{T}(0)}$. So we can assume that the hidden loci are in linkage equilibrium with each other and with the observed loci. Restricting to this submanifold, the observed selection field is $\bar{\nabla} \frac{1}{2} \overline{m^T}$ by Prop. 1.7 and the observed recombination fields are of the form (3.1) with r^S replaced by $r^{S,T}$, i by i_T, etc., by Prop. 6. This is the nice case in which the genetic background has no observable effect.

Recombination Rates Constant; Birth and Death Rates Show $T \vee \tilde{T}$
<u>Type Epistasis</u>: This means that $b = b^T + b^{\tilde{T}}$, $d = d^T + d^{\tilde{T}}$. On $\Lambda_{T\vee\tilde{T}}$ the observed selection field will be $\bar{\nabla} \frac{1}{2} \overline{m^T}$ by Prop. 1.7 again and for the observed recombination fields there will be one term of the form (3.1) with r^S replaced by $r^{S,T}$ and b by b^T, etc. plus a term contributed by $b^{\tilde{T}}$. The latter term will be of the form (3.2) with r^S replaced by $r^{S,T}$, i replaced by i_T, etc. and with b replaced by $\mathbb{E}(b^{\tilde{T}})$. So on $\Lambda_{T\vee\tilde{T}}$ the effect of genetic background will appear by varying the strength of this added recombination term. However, $\Lambda_{T\vee\tilde{T}}$ is not an invariant submanifold for recombination in Δ and so we may move off it. Once we do the observed loci are no longer independent of the hidden loci, observed position effects will appear unless $b^{\tilde{T}} = 0$. In the observed selection field new terms will appear depending on the contributions to fitness of the hidden loci, $m^{\tilde{T}}$, and the distance from $\Lambda_{T\vee\tilde{T}}$ as measured by the functions d_i^T (see again Prop. 1.7).

4. Position Effects.

For simplicity we will assume that r_{ij}^S is completely symmetric and focus on b_{ij}, which we will assume are positive for all $i,j \in I$. The recombination vectorfields are given by (c.f. I.(7.1) and I.(7.4)):

$$(4.1) \qquad R^S = \sum_{i,j} r_{ij}^S (b_{ij}p_ip_j - b_{\bar{i}\bar{j}}p_{\bar{i}}p_{\bar{j}})\partial_i$$

$$= (1/4) \sum_{i,j} r_{ij}^S (b_{ij}p_ip_j - b_{\bar{i}\bar{j}}p_{\bar{i}}p_{\bar{j}}) \bar{\nabla} L_{ij}^S$$

with $\bar{i} = i_S j_{\tilde{S}}$ and $\bar{j} = j_S i_{\tilde{S}}$. The recombination field is still $-R$ with $R = \Sigma\{R^S : S \subset L\}$.

The conditions of Theorem 2.1 still hold for the general recombination field. Furthermore, we can mimic Prop. 2.4 by defining:

$$(4.2) \qquad L_{ij}^{S,b}(p) = \ell n(b_{ij}p_ip_j / b_{\bar{i}\bar{j}}p_{\bar{i}}p_{\bar{j}})$$

$$= \ell n(b_{ij}p_ip_j) - \ell n(b_{\bar{i}\bar{j}}p_{\bar{i}}p_{\bar{j}})$$

$$= L_{ij}^S(p) + \ell n(b_{ij}/b_{\bar{i}\bar{j}}).$$

$$(4.3) \qquad Q_{ij}^{S,b}(p) = Q(b_{ij}p_ip_j, b_{\bar{i}\bar{j}}p_{\bar{i}}p_{\bar{j}}).$$

Since $L_{ij}^{S,b}$ and L_{ij}^S differ by a constant, they have the same gradient. So $L^{S,b} \bar{\nabla} L^S = \frac{1}{2} \bar{\nabla}(L^{S,b})^2$. From this follows the analogue of (2.3):

$$(4.4) \qquad R^S = (1/8) \sum_{i,j} r_{ij}^S Q_{ij}^{S,b} \bar{\nabla}(L_{ij}^{S,b})^2.$$

But if we take the inner product with the gradient of entropy we run into trouble because

$$(\bar{\nabla}_p (L_{ij}^{S,b})^2, \bar{\nabla}_p H)_p = -2(L_{ij}^{S,b})(L_{ij}^S)$$

and the right side need not be positive. However, there is a special case where we can generalize Theorem 2.5:

<u>1 Theorem:</u> The following conditions on $b \in R^{I \times I}$ are equivalent and define the condition: b_{ij} shows <u>simple position effects</u>.

(a) There exists $q \in R^I$ with $q_i > 0$ for all i such that

(4.5)
$$b_{ij}/b_{\bar{i}\bar{j}} = q_i q_j / q_{\bar{i}} q_{\bar{j}}$$

for all $i,j \in I$ and $S \subset L$, where $\bar{i} = i_S j_{\tilde{S}}$ and $\bar{j} = j_S i_{\tilde{S}}$.

(b) There exists $p \in \mathring{\Delta}$ such that the functions $\{L_{ij}^{S,b}: i,j \in I \ S \subset L\}$ vanish simultaneously at p.

(c) There is a leaf of the transverse foliation $\bar{\mathcal{J}}$ such that R vanishes exactly on the leaf for all positive choices of the r_{ij}^S's.

(d) The vectorfields $\{\bar{\nabla}(L_{ij}^{S,b})^2: i,j \in I, S \subset L\}$ are coherent in the sense that if $\lambda_{ij}^S \geq 0$ for i,j,S and

$$\sum_{i,j,S} \lambda_{ij}^S \bar{\nabla}(L_{ij}^{S,b})^2 = 0 \quad \text{at } p \in \mathring{\Delta}$$

then $\lambda_{ij}^S L_{ij}^{S,b} = 0$ for all i j,S at p.

If b_{ij} shows simple position effects define, for q satisfying condition (a):

(4.6)
$$H^q(p) = H(p) - E^{\ln P}(p) = H(p) - \overline{\ln q}$$

$$= -\sum p_i \ln p_i - \sum p_i \ln q_i = -\sum p_i \ln p_i q_i$$

$$= -\frac{1}{2}\sum p_i p_j \ln(p_i q_i p_j q_j).$$

The following equation holds on $\overset{\circ}{\Delta}$:

4.7) $$(-R, \bar{\nabla}_p H^q)_p = (1/4) \sum_{i,j,S} r^S_{ij} Q^{S,b}_{ij} (L^{S,b}_{ij})^2.$$

The sum is nonnegative and vanishes exactly when R vanishes and this is on the leaf of $\bar{\mathcal{J}}$ defined by the points on which $L^{S,b}_{ij}$ all vanish, i.e. points at which the probability distribution p^q is in $\overset{\circ}{\Lambda}$ where:

4.8) $$p^q(i) = P_i q_i / \Sigma_j \ P_j q_j.$$

Proof: We begin by assuming (a) and prove (4.7).

From I.4.13 we have:

$$L^{S\,b}_{ij} = L^S_{ij} + \ell n \ q_i - \ell n \ q_{\bar{i}} - \ell n \ q_{\bar{j}} + \ell n \ q_j$$

$$= -(\bar{\nabla}_p H, \bar{\nabla}_p L^S_{ij})_p + (\bar{\nabla}_p E^{\ell n \ q}, \bar{\nabla}_p L^S_{ij})_p.$$

Since $\bar{\nabla} L^{S,b} = \bar{\nabla} L^S$ we have

4.9) $$(\bar{\nabla}_p H^q, \bar{\nabla}_p L^{S,b}_{ij})_p = -L^{S,b}_{ij}.$$

(4.7) now follows just as (2.8) did. Just as in Theorem 2.5, the sum is positive and vanishes where R does which is when all of the $L^{S,b}_{ij}$'s do. Defining for $p \in \overset{\circ}{\Delta}$ the vector x by $x_i = p_i q_i$, we see that $L^{S,b}_{ij}(p) = L^S_{ij}(x)$ and since L is homogeneous of degree zero we can normalize to get

4.10) $$L^{S,b}_{ij}(p) = L^S_{ij}(p^q).$$

Thus, the $L^{S,b}_{ij}$'s vanish at p iff the L^S_{ij}'s vanish at p^q iff $p^q \in \overset{\circ}{\Lambda}$.

The solutions of $\{L_{ij}^{S,b} = 0\}$ are the same as the solutions of $\{L_{ij}^{S} = -L_{ij}^{S}(q)\}$ which defines a leaf of $\bar{\mathcal{J}}$ if the constants are consistent i.e. if any solution exists. Solutions exist because $p_i = q_i^{-1}/\Sigma_j \; q_j^{-1}$ is a solution with $p^q(i) = 1/n$. This proves (4.8) and also shows that (a) implies (c).

Also, (a) implies (d) because by (4.9)

$$(-\bar{\nabla}_p H^q, \; \sum \lambda_{ij}^{S} \bar{\nabla}_p (L_{ij}^{S,b})^2)_p = 2 \sum \lambda_{ij}^{S} (L_{ij}^{S,b})^2$$

and the right vanishes iff $\lambda_{ij}^{S} L_{ij}^{S,b} = 0$ for all i,j,S.

(c) implies, by varying the r_{ij}^{S}'s, that all of the $L_{ij}^{S,b}$'s vanish on the specified leaf. This implies (b).

If (b) holds and $L_{ij}^{S,b}(p^0) = 0$ for all i,j,S then (4.5) holds with $q_i = (p_i^0)^{-1}$ and this implies (a). Thus, (a) - (c) are equivalent and (a) implies (d).

Finally, if (d) holds then consider the vectorfield $-R$ on $\mathring{\Delta}$. This vectorfield extends to Δ using the original definition of (4.1). If $p_i = 0$ the coefficient of ∂_i in $-R$ is positive and if $p_i = 1$ the coefficient of ∂_i is netative. So on the Δ-closure of any leaf of $\mathcal{D}$, $-R$ is a vectorfield on a convex cell pointing inward at the boundary. So by one of the equivalents of the Brouwer Fixed Point Theorem it vanishes somewhere in the open cell. Applying (d) at this point with $\lambda_{ij}^{S} = r_{ij}^{S} Q_{ij}^{S,b} > 0$ we get that the functions $\{L_{ij}^{S,b}\}$ vanish simultaneously at some point. This is (c). $\hspace{2cm}$ QED

Remarks: (a) We can obtain a Lyapunov function analogous to $\hat{H}$ but in a rather noncomputable fashion. Let Λ_b denote the leaf at which the $L_{ij}^{S,b}$'s all vanish. Since $E|\Lambda_b: \Lambda_b \to \Pi_\alpha \mathring{\Lambda}^\alpha$ is a diffeomorphism we can define

(4.11) $$\hat{H}^b = H^q - H^q \cdot (E|\Lambda_b)^{-1} \cdot E.$$

Then $\hat{H}^b$ equals 0 on Λ_b and differs with H^q by a function which factors through E. So (4.7) holds with $\bar{\nabla}H^q$ replaced by $\bar{\nabla}\hat{H}^b$. Furthermore, $\bar{\nabla}\hat{H}^b$ vanishes on Λ_b just as in Lemma II.1.7, i.e. $\bar{\nabla}H^q$ is tangent to Λ_b because at each point of Λ_b the restriction of H^q to the perpendicular leaf of $\mathcal{D}$ takes on its maximum value by the same argument as Prop. II.1.6.

The projection map $(E|\Lambda_b)^{-1} \cdot E^{(0)} : \overset{\circ}{\Delta} \to \Lambda_b$ is rather difficult to compute. One can only say that by Prop. II.2.11(b) its tangent map at points of Λ_b is the orthogonal projection of $T_p\Delta$ on $T_p\Lambda_b$.

(b) The choice of q is not uniquely defined by (4.5). But if r_i is the ratio of two different choices for q_i then $r_i > 0$, and $r \in R^I$ such that $r_i r_j$ is completely symmetric. If we normalize r to get an element of $\overset{\circ}{\Delta}$, i.e. divide by $|r| = \Sigma_j r_j$, Prop. II.2.14(a) implies that the resulting distribution lies in $\overset{\circ}{\Lambda}$, i.e. $r_i = C\Pi_\alpha r_{i_\alpha}^\alpha$. Conversely, if we multiply a solution q_i of (4.5) by Cr_i for $r \in \overset{\circ}{\Lambda}$ we get another solution of (4.5). p^q and H^q depend on the choice of q, but for differing choices of q the corresponding H^q's differ by $E^{\ell n \, r}$. This factors through E since r is a multiple of a distribution in $\overset{\circ}{\Lambda}$. In particular, the differences are canceled by normalizing and $\hat{H}^b$ depends only on b_{ij}, which is why we did not denote it $\hat{H}^q$.

(c) Another way of describing Λ_b is to say it consists of all $p \in \overset{\circ}{\Delta}$ such that $p_i p_j b_{ij}$ is completely symmetric if any such p's exist. b_{ij} shows simple position effects when such p's exist (this is part

(c) of the definition). This suggests a function space interpretation for simple position effects. Let $Sym \subset R^{I \times I}$ be the subspace
of symmetric functions and $Sym^* \subset Sym$ be the subspace of completely
symmetric functions. Let $\mathcal{L}_{L+L} \subset R^{I \times I}$ consisting of functions of
ij which are sums of functions of i alone and of j alone. In
the notation of Sec. II.2 this is the set of functions whose carriers
are contained in $L + L = L \times \{\emptyset\} \vee \{\emptyset\} \times L$. $Sym \cap \mathcal{L}_{L+L}$ defines
Sym_{L+L} which is isomorphic to R^I as each element is of the form
$u_i + u_j$ for some $u \in R^I$. Now ℓn b, defined by $(\ell n\ b)_{ij} = \ell n\ b_{ij}$, is
always a member of Sym. If it lies in Sym* there are no position
effects. If it lies in $Sym^* + Sym_{L+L}$ the position effects are
simple.

I don't know or any way of detecting whether or not position
effects are simple. The obvious approach is to use the above function
space interpretation. If b_{ij} is not completely symmetric, i.e.
$\ell n\ b \in Sym^*$, one tries to project to Sym* and check whether the error
lies in Sym_{L+L}. In the notation of Thm. II.2.9, we define
$S: R^{I \times I} \to Sym^*$ by

(4.12) $$S(n) = 2^{-\ell} \sum \{T_Q(n) : Q \subset L\}.$$

With respect to the standard Euclidean metric (,) on $R^{I \times I}$, S is the
orthogonal projection because the T_Q's are isometries, see Prop. II.
2.14. I conjectured in vain that b_{ij} has simple position effects iff
$\ell n\ b - S(\ell n\ b) \in Sym_{L+L}$. While clearly sufficient this condition is
not necessary because S does not preserve Sym_{L+L}.

I am also unable to concoct any convincing biological reason
why position effects, if they occur, are likely to be simple. For

example, I see no reason that the observed position effects of the previous sections need be simple. The simplicity is exclusively in the fact that we can analyze the model. Even if nonsimple position effects occur, we can apply the Brouwer Theorem to $-R$ on each leaf of the fibre foliation. There exist points of each leaf on which R vanishes, because each leaf is an open convex call and $-R$ is directed inwards at the boundary. However, the failure of condition (c) means that the equilibria for R depend on the recombination rates, r_{ij}^s, as well as on b_{ij}. In the case of simple position effects the set of equilibria of R, namely Λ_b, depends only on b. If there are no position effects then the set of equilibria of R is $\overset{\circ}{\Lambda}$ and so is independent of b, too.

The portrait painted by Theorem 1 is structurally stable. This means that if the nonsimple position effects are small enough, i.e. $\ell n\ b$ is close enough to the subspace $Sym^* + Sym_{L+L}$, then there is a smooth manifold of equilibria, $\tilde{\Lambda}$, for R C^∞ close to $\Lambda_{\bar{b}}$ where $\ell n\ \bar{b}$ is the projection of $\ell n\ b$ on $Sym^* + Sym_{L+L}$. This follows from a parametrized version of the structural stability theorem for linearly stable equilibria. Apply the theorem to $-R$ on each leaf of the foliation $\mathfrak{D}$. Furthermore, $\tilde{\Lambda}$ is globally attracting for $-R$ on each leaf, again assuming $\ell n\ b$ is close enough to $Sym^* + Sym_{L+L}$.

One case where these results always apply is to the two-locus-two-allele model ($n_1 = n_2 = \ell = 2$). There is not too much room in such a model so the position effect, if it occurs, is simple.

<u>2 Corollary</u>: Consider a two-locus-two-allele model with $\ell = 2$, $I_1 = \{A,a\}$ and $I_2 = \{B,b\}$. We number the gametes: $1 = AB$, $2 = Ab$, $3 = aB$, $4 = ab$. The only nonzero functions L are:

(4.13) $$L_{14}^1 = -L_{23}^1 = \ell n \ p_1 p_4 / p_2 p_3.$$

Assume all of the birth-rates are positive and let ρ denote the ratio of the "coupling" to "repulsion" birth rates, i.e. $\rho = b_{14}/b_{23}$. If $\rho = 1$ there are no position effects. If $\rho \neq 1$ then b shows only simple position effects.

The gene frequencies $p_A = p_1 + p_2$ and $p_B = p_1 + p_3$ are invariant under the recombination field and on each cell defined by a choice of p_A and p_B the recombination field approaches the point satisfying $L_{14}^1 = -\ell n \ \rho$. If $\rho = 1$, this is the distribution with $p_1 = p_A p_B$, etc.

Proof: Let $q_1 = \rho$, $q_2 = q_3 = q_4 = 1$. (4.5) is satisfied by inspection.   QED

5. Mutation.

The mutation field is defined by (cf. I.(8.1) and I.(8.2)):

$$N = \sum_{i,j} p_j N_{ji} \partial_i$$

(5.1)

$$N_{ji} = \begin{cases} n_{ji} & i \neq j \\ -n_{i*} & i = j \end{cases}$$

where $n_{i*} = \Sigma \ n_{ij}$ summed on all $j \neq i$.

Notice that the coefficient of ∂_i is positive if $p_i = 0$ and that the sum of the coefficients is 0. So N is a vectorfield on Δ pointing inward on the boundary. On the other hand, the formula (5.1) with p_j replaced by x_j extends the definition of N to a

linear vectorfield on all of R^I which we will also denote by N.
Define $(R^I)_0 = \{x \in R^I: \Sigma x_i = 0\}$. Recall that this subspace is the
tangent space at p of the manifold $\mathring{\Delta}$, $T_p\mathring{\Delta}$.

1 Theorem: Assume $n_{ji} > 0$ whenever $i \neq j$. On Δ the mutation field
has a unique equilibrium point, q, which lies in $\mathring{\Delta}$. The change of
variable $x = p - q$ translates N to its restriction to $(R^I)_0$. It is
the linear differential equation associated with the matrix N which
maps $(R^I)_0$ to itself. Let $\rho = \max_i n_{i*}$. The set of eigenvalues of
$N|(R^I)_0$, i.e. the spectrum of $N|(R^I)_0$, lies in the disc:

$$\text{Spec}(N|(R^I)_0) \subset \{z \in \mathbb{C}: |z+\rho| \leq \rho\} - \{0\},$$

where $\mathbb{C}$ is the complex plane. Since all of the eigenvalues have
negative real parts, 0 is a globally, asymptotically stable equili-
brium for $N|(R^I)_0$ and q is a globally, asymptotically stable equili-
brium for N on Δ. If we define the rate constant $\rho(N)$ by

$$-\rho(N) = \max\{\text{Real part } z: z \in \text{Spec}(N|R^I)_0)\}$$

then $\rho(N)$ measures the rate of approach to equilibrium.

(5.2) $0 < \rho(N) \leq 2\rho = 2 \max_i n_{i*}.$

Proof: Regarding N as a vectorfield on R^I or as a matrix, an
equilibrium q is a vector satisfying $qN = 0$, i.e. $\Sigma_j q_j N_{ji} = 0$ for
all i. Also if 1 denotes the vector all of whose entries are 1,
then $N1 = 0$, i.e. $\Sigma_i N_{ji} = 0$ for all j.

Now let λ be greater than the positive number ρ. The
matrix $N + \lambda I$ is a positive matrix and so the machinery of the
Frobenius theory of positive matrices applies. For a nice exposition

of this theory see the Appendix of Karlin's book [18]. If P is a positive matrix, i.e. a matrix with all entries positive, then there are vectors r and ℓ and a positive number a satisfying:

(1) r and ℓ are positive vectors and are right and left eigenvectors for P with eigenvalue a, i.e.:

$$Pr = ar \quad \text{and} \quad \ell P = a\ell.$$

(2) The multiples of ℓ are the only left eigenvectors of P associated with the eigenvalue a and there are no other nonnegative left eigenvectors associated with any eigenvalue. Similarly for r and the right eigenvectors.

(3) Let $[r]^{\perp}$ consist of all vectors x such that $(x,r) = 0$ where (,) is the usual Euclidean inner product. $x \in [r]^{\perp}$ implies $xP \in [r]^{\perp}$ and the spectrum of the restriction $P|[r]^{\perp}$ is the spectrum of P with a removed. This spectrum is contained in the open disc of radius a:

$$\text{Spec } P|[r]^{\perp} \subset \{z \in \mathbb{C}: |z| < a\}.$$

Now we apply all this to $P = N + \lambda I$. Since $N1 = 0$ $P1 = \lambda 1$ and so by (2) $a = \lambda$ and we can let $r = 1$. By (1) there is a positive left eigenvector ℓ and we can normalize to get $q_i = \ell_i / \Sigma_j \ell_j$. So by (2) q is the unique left eigenvector of $N + \lambda I$ in Δ and has eigenvalue λ. So $qN = 0$. Since $q_i > 0$ for all i, $q \in \overset{\circ}{\Delta}$.

The translation result is clear and since $(R^I)_0 = [1]^{\perp}$, (3) implies that the spectrum of $(N + \lambda I)|(R^I)_0$ is contained in the open disc of radius λ. Subtracting λI just translates the spectrum and so we have

$$\text{Spec } N \mid (R^I)_0 \subset \{z \in \mathbb{C} \colon |z + \lambda| < \lambda\}.$$

The intersection of these open discs as λ approaches ρ from above is $\{z \in \mathbb{C} \colon |z + \rho| \leq \rho\} - \{0\}$. The estimates on $\rho(N)$ are then clear and the interpretation in terms of stability comes from the standard theory of linear differential equations with constant coefficients, see e.g. [15].   QED

Remark: For the Frobenius theory to apply to a matrix P it is necessary that all the entries be nonnegative but they needn't all be positive. It suffices that for some power of P all entries be positive. This means that we need not assume that $n_{ji} > 0$ for all $i \neq j$. It is sufficient to assume: (1) $n_{ji} \geq 0$ for all $i \neq j$ and (2) for every ordered pair (j,i) with $i \neq j$ we can get from j to i by a sequence of mutations, i.e. there is a sequence from j to i of distinct elements of I with $n_{k\ell} > 0$ for k, ℓ successive members of the sequence. See again the Appendix of [18] or for a nice graph theoretic treatment of this problem, Demetrius [7].

2 Lemma: Let $A = (a_{ij})$ be a square matrix of corank $= 1$ meaning that 0 is an eigenvalue for A and the associated left and right eigenspaces are one-dimensional. Thus, there are nonzero vectors r and ℓ, unique up to constant multiple, satisfying

$$Ar = 0 \qquad \text{and} \qquad \ell A = 0.$$

Let M be the associated cofactor matrix for A, i.e. $M_{ij} = (-1)^{i+j}$ times the ji minor of A. Then there exists a nonzero constant K such that

$$M_{ij} = Kr_i \ell_j.$$

Proof: By Cramer's rule AM = MA = det(A)I and this equals 0 because A is singular. But because the corank of A is 1, some (n-1) × (n-1) minor is nonzero and so M is not the zero matrix. Since AM = 0 each column of M is a multiple of r. So $M_{ij} = r_i K_j$ for some constants K_j not all zero. Since MA = 0 each row of M is a multiple of ℓ. Now if $r_{i_0} \neq 0$ then we can write the i_0 row of M as $Kr_{i_0}\ell_j$ and so get $K_j = K\ell_j$.   QED

3 Corollary: For the mutation field matrix N let M_{ii} be the ii minor. Then

$$q_i = M_{ii} / \Sigma_j \, M_{jj}.$$

Proof: We apply Lemma 2 with A = N. In the proof of Thm. 1, we showed that the eigenspaces of N + λI associated with λ is one-dimensional. So the eigenspaces of N associated with 0 are one-dimensional and we can choose ℓ = q and r = 1. Lemma 2 then says that $M_{ij} = Kq_j$ for some nonzero constant K. In particular, $M_{ii} = Kq_i$ and $\Sigma_j \, M_{jj} = K$.   QED

There is an important special case where the equilibrium is obvious and where there is a simple Lyapunov function for the mutation field. This is when the forward and backward mutation rates are the same.

4 Theorem: Suppose $n_{ij} = n_{ji}$ for all i,j distinct in I. The center of the simplex q is the equilibrium of the mtation field. The function f: $\Delta \rightarrow$ R:

$$(5.3) \qquad f(p) = -\sum (p_i - q_i)^2$$

is a Lyapunov function for N on Δ.

Proof: In this case the matrix N is symmetric. The differential equation for mutation extended to R^I is

$$\frac{dx}{dt} = xN.$$

With $(\ ,\ )$ the usual inner product on R^I, we have

$$(5.4) \qquad \frac{d\,\frac{1}{2}(x,x)}{dt} = (xN, x).$$

On the invariant subspace $(R^I)_0$ the eigenvalues of N have negative real parts by Thm. 1 and so by symmetry are negative. This implies that the quadratic function (xN,x) is negative definite on $(R^I)_0$, i.e.

$$(5.5) \qquad (xN,x) < 0 \qquad \text{if} \quad x \in (R^I)_0 \text{ and } x \neq 0.$$

Since $N1 = 0$ symmetry implies $1N = 0$ and so the equilibrium q is the vector 1 normalized to lie in Δ, i.e. $q_i = 1/n$ where n is the number of elements in I. Since $qN = 0$ we have for p in Δ:

$$(5.6) \qquad \frac{d\,\frac{1}{2}(p-q,p-q)}{dt} = (pN, p - q) = ((p - q)N,\ p - q).$$

By (5.5) this is negative unless $p = q$. This proves (5.3) since $\frac{1}{2}(p-q,p-q) = -\frac{1}{2}f(p)$. \hfill QED

Remark: In this case the mutation field is the gradient of the quadratic function $\frac{1}{2}((p-q)N, p-q)$ but it is the gradient with respect to the usual inner product not with respect to the Shahshahani metric.

For the multilocus model with $I = \Pi_\alpha I_\alpha$, the mutation field for the α locus on Δ_α is:

$$\bar{N}^\alpha = \sum_{i_\alpha, j_\alpha} p^\alpha_{j_\alpha} \bar{N}^\alpha_{j_\alpha i_\alpha} \partial_{i_\alpha}$$

(5.7)

$$\bar{N}^\alpha_{j_\alpha i_\alpha} = \begin{cases} n^\alpha_{j_\alpha i_\alpha} & j_\alpha \neq i_\alpha \\[2mm] -n^\alpha_{i_{\alpha *}} & j_\alpha = i_\alpha \end{cases}$$

where $n^\alpha_{i_{\alpha *}} = \Sigma\, n^\alpha_{i_\alpha j_\alpha}$, summed on all $j_\alpha \neq i_\alpha$.

On Δ the partial mutation field corresponding to mutation at the α locus is:

$$N^\alpha = \sum_{i,j} p_j N^\alpha_{ji} \partial_i$$

(5.8)

$$N^\alpha_{ji} = \bar{N}^\alpha_{j_\alpha i_\alpha} \delta_{\tilde{j}_{\tilde\alpha} \tilde{i}_{\tilde\alpha}} .$$

The Kronecker delta notation means that $N^\alpha_{ji} = \bar{N}^\alpha_{j_\alpha i_\alpha}$ if $i = j$ at all loci other than α and $N^\alpha_{ji} = 0$ otherwise.

Finally, the full mutation field is the sum:

$$N = \sum_\alpha N^\alpha = \sum_{i,j} p_j N_{ji} \partial_i$$

(5.9)

$$N_{ji} = \sum_\alpha N^\alpha_{ji}.$$

Extending $\bar{N}^\alpha$ to a linear vectorfield on R^{I_α}, and N^α, N to linear vectorfields on R^I, we can identify these linear vectorfields with the corresponding linear maps and with the matrices operating on the right, e.g. $N: R^I \to R^I$ by $N(x)_i = \Sigma_j x_j N_{ji}$.

<u>5 Theorem:</u> For $\alpha, \beta \in L = \{1, \dots, \ell\}$ the following diagram commutes (meaning the two composed maps are equal):

$$
\begin{array}{ccc}
R^I & \xrightarrow{\ E^\beta\ } & R^{I_\beta} \\
\uparrow{\scriptstyle N^\alpha} & & \uparrow{\scriptstyle \bar{N}^\beta \delta_{\alpha\beta}} \\
R^I & \xrightarrow{\ E^\beta\ } & R^{I_\beta}
\end{array}
$$

so that if $\alpha = \beta$ the vertical map on the right is $\bar{N}^\alpha$ and if $\alpha \neq \beta$ the vertical map is 0.

The following diagram commutes:

$$
\begin{array}{ccc}
R^I & \xrightarrow{\ E\ } & \Pi_\alpha R^{I_\alpha} \\
\uparrow{\scriptstyle N} & & \uparrow{\scriptstyle \Pi_\alpha \bar{N}^\alpha} \\
R^I & \xrightarrow{\ E\ } & \Pi_\alpha R^{I_\alpha}
\end{array} \quad .
$$

Thus, the vectorfields N and $\Pi_\alpha \bar{N}^\alpha$ are E related meaning that E maps the vector N at p to the vector $\Pi_\alpha \bar{N}^\alpha$ at $E(p)$.

Assume that $n^\alpha_{j_\alpha i_\alpha} > 0$ whenever $i_\alpha \neq j_\alpha$, for all α. Let $q^\alpha \in \mathring{\Delta}_\alpha$ be the equilibrium for the vectorfield $\bar{N}^\alpha$. The unique globally asymptotically stable equilibrium point, q, of the mutation field N on Δ satisfies:

(5.10)
$$
q_i = \Pi_\alpha q^\alpha_{i_\alpha}
$$

and so $q \in \mathring{\Lambda} \subset \mathring{\Delta}$.

Defining the rate constants $\rho(N)$ and $\rho(\bar{N}^\alpha)$ as in Thm. 1, we have

(5.11) $0 < \rho(N) \le \min_\alpha \rho(\bar{N}^\alpha) \le 2 \min_\alpha \max_i n^\alpha_{i_\alpha i_{\alpha^*}}.$

<u>Proof</u>:

(5.12) $N^\alpha(x)_i = \sum_j x_j N^\alpha_{ji} = \sum_{j_\alpha} x_{j_\alpha i_{\tilde{\alpha}}} \bar{N}^\alpha_{j_\alpha i_\alpha}$

i.e. the only nonzero terms in the first sum are those where j
agrees with i at the loci other than α. Now applying E^α amounts
to summing over the $i_{\tilde{\alpha}}$ indices. So we have

$$E^\alpha(N^\alpha(x))_{i_\alpha} = \sum_{j_\alpha} E^\alpha(x)_{j_\alpha} \bar{N}^\alpha_{j_\alpha i_\alpha} = N^\alpha(E^\alpha(x))_{i_\alpha}.$$

On the other hand, applying E^β with $\beta \ne \alpha$ includes summing
over the i_α indices. The row sums of N^α are all zero, i.e. $N^\alpha_{j_{\alpha^*}} = 0$
for all j_α. So $E^\beta(\bar{N}^\alpha(x)) = 0$ if $\beta \ne \alpha$. Thus the first diagram
commutes, and so for each α does the following

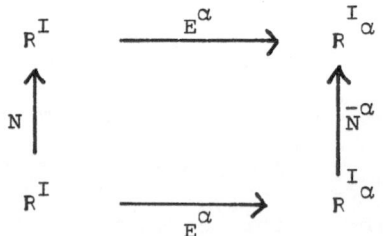

This implies commutativity of the second.

   Now if we apply (5.12) with $x = q$ defined by (5.10) we get

$$N^\alpha(q)_i = (\sum q^\alpha_{j_\alpha} \bar{N}^\alpha_{j_\alpha i_\alpha}) \Pi_{B \ne \alpha} q^\beta_{i_\beta} = 0.$$

So q is an equilibrium for each N^α and so for N. But by Thm. 1

and the Remark thereafter N has a unique equilibrium and it is globally, asymptotically stable. So q of (5.10) is it.

Finally, the rate constant estimate (5.11) follows from:

$$\cup_\alpha \; \mathrm{Spec}\,(\bar{N}^\alpha |\, (R^{I_\alpha})_0) \subset \mathrm{Spec}\,(N|\,(R^I)_0).$$

To prove this let $x \in (R^{I_\alpha})_0$ left eigenvector for N^α with eigenvalue z, i.e. $\Sigma_{j_\alpha} x_{j_\alpha} \bar{N}^\alpha_{j_\alpha i_\alpha} = zx_{i_\alpha}$. Define $\bar{x} \in (R^I)_0$ by $\bar{x}_i = x_{i_\alpha} \Pi_{\beta \neq \alpha} q^\beta_{i_\beta}$ and apply (5.12):

$$\sum_j \bar{x}_j N^\alpha_{ji} = \left(\sum_{j_\alpha} x_{j_\alpha} \bar{N}^\alpha_{j_\alpha i_\alpha} \right) \Pi_{\beta \neq \alpha} q^\beta_{i_\beta} = z\bar{x}_i$$

$$\sum_j \bar{x}_j N^\mu_{ji} = x_{i_\alpha} \left(\sum_{j_\mu} q^\mu_{j_\mu} \bar{N}^\mu_{j_\mu i_\mu} \right) \Pi_{\beta \neq \alpha, \mu} q^\beta_{i_\beta} = 0 \qquad (\mu \neq \alpha).$$

So $\bar{x}$ is a left eigenvector for $\bar{N}$ with eigenvalue z. $\qquad$ QED

Actually we can say much more than just that q lies in the Wright manifold.

6 Addendum: The mutation fields N^α and N are all tangent to the Wright manifold $\stackrel{\circ}{\Lambda}$ at all points of $\stackrel{\circ}{\Lambda}$. So $\stackrel{\circ}{\Lambda}$ is an invariant manifold for the associated flows.

Proof: This can be proved by direct computation. N^α is tangent to $\stackrel{\circ}{\Lambda}$ iff N^α_i/p_i as a function of i lies in $\mathcal{L}^{(0)}$ for $p \in \stackrel{\circ}{\Lambda}$ (cf. Addendum II.1.3(2)). If $p \in \stackrel{\circ}{\Lambda}$ this is $\Sigma_{j_\alpha} p_{j_\alpha} \bar{N}^\alpha_{j_\alpha i_\alpha}/p_{i_\alpha}$ which depends only on the α locus. $\qquad$ QED

Remark: It is crucial for this result not only that mutation occur independently at separate loci, but also that the mutation rates at the α locus depend only on the alleles at that locus. It is no longer true when genes at one locus influence the mutation rate at others.

Addendum 6 could also be proved using the following result which is related to Shahshahani's Prop. 3.3 [28]:

7 Proposition: (a) The vectorfields N^α and N^β on R^I commute, i.e. the Lie bracket $[N^\alpha, N^\beta] = 0$.

(b) Let $S \subset L$ and assume that the birth rates b_{ij} and recombination rates r_{ij}^S are genotype independent, i.e. $b_{ij} = b$ and $r_{ij}^S = r^S$ for all $i, j \in I$. In that case, the mutation fields N^α and the recombination fields R^S commute, i.e. $[N^\alpha, R^S] = 0$ for all α, S. So the total mutation field N and the total recombination field $R = \Sigma_S R^S$ commute.

Proof: (a): Commuting of linear vectorfields is the same as the commuting of the associated matrices. Since any matrix commutes with itself, we can assume $\alpha \neq \beta$ and show that $N^\alpha N^\beta = N^\beta N^\alpha$. It is easy to check that:

$$(N^\alpha N^\beta)_{ij} = \bar{N}^\alpha_{i_\alpha j_\alpha} \bar{N}^\beta_{i_\beta j_\beta} \delta^i{}_{\{\alpha,\beta\}} j_{\{\alpha,\beta\}}.$$

This is clearly symmetric in α and β.

(b): Define

$$N^\alpha_i = \sum_j p_j N^\alpha_{ji} = \sum_j p_j \bar{N}^\alpha_{j_\alpha i_\alpha} \delta_{j_{\tilde\alpha} i_{\tilde\alpha}} = \sum_{j_\alpha} p_{j_\alpha i_{\tilde\alpha}} \bar{N}^\alpha_{j_\alpha i_\alpha}.$$

So that $N^\alpha = \Sigma_i N^\alpha_i \partial_i$. Since $\bar{N}^\alpha_{j_{\alpha^*}} = 0$,

$$(5.13) \qquad \sum_{j_S} N^{\alpha}_{j_S i_{\widetilde{S}}} = 0 \quad \text{if} \quad \alpha \in S \ (i_{\widetilde{S}} \in I_{\widetilde{S}} \text{ fixed}).$$

Now if $\alpha \in S$, let $T = S - \{\alpha\}$. Then it is also clear that

$$(5.14) \qquad \sum_{j_{\widetilde{S}}} N^{\alpha}_{i_S j_{\widetilde{S}}} = \sum_{j_\alpha} P_{j_\alpha i_T} \bar{N}^{\alpha}_{j_\alpha i_\alpha} = \sum_{j_S} P_{j_S} \bar{N}^{\alpha}_{j_\alpha i_\alpha} \delta_{j_T i_T}.$$

Now to show $[N^{\alpha}, R^S] = 0$, we must show that
$\Sigma_j \ d^S_j \partial_j N^{\alpha}_i = \Sigma_j \ N^{\alpha}_j \partial_j d^S_i$ for all i, on Δ (cf. (3.2)). Since $R^S = R^{\widetilde{S}}$ we can assume $\alpha \in S$.

$$\sum_j N^{\alpha}_j \partial_j d^S_i = \sum_j N^{\alpha}_j (\delta_{ij} - P_{i_{\widetilde{S}}} \delta_{j_S i_S} - P_{i_S} \delta_{j_{\widetilde{S}} i_{\widetilde{S}}})$$

$$= N^{\alpha}_i - P_{i_{\widetilde{S}}} \sum_{j_{\widetilde{S}}} N^{\alpha}_{i_S j_{\widetilde{S}}} - P_{i_S} \sum_{j_S} N^{\alpha}_{j_S i_{\widetilde{S}}}$$

$$= N^{\alpha}_i - P_{i_{\widetilde{S}}} \sum_{j_{\widetilde{S}}} N^{\alpha}_{i_S j_{\widetilde{S}}} \qquad \text{(by (5.13)).}$$

On the other hand,

$$\sum_j d^S_j \partial_j N^{\alpha}_i = \sum_j (P_j - P_{j_S} P_{j_{\widetilde{S}}}) N^{\alpha}_{ji}$$

$$= N^{\alpha}_i - \sum_j P_{j_S} P_{j_{\widetilde{S}}} \bar{N}^{\alpha}_{j_\alpha i_\alpha} \delta_{j_{\widetilde{\alpha}} i_{\widetilde{\alpha}}}.$$

Now $\delta_{j_{\widetilde{\alpha}} i_{\widetilde{\alpha}}} = \delta_{j_{\widetilde{S}} i_{\widetilde{S}}} \delta_{j_T i_T}$ and so summing first on $j_{\widetilde{S}}$ and then applying (5.14) we have

$$\sum_{j} d^S_j \partial_j N^{\alpha}_i = N^{\alpha}_i - \sum_{j_S} p_{j_S} p_{i_{\underset{\sim}{S}}} \bar{N}^{\alpha}_{j_{\alpha}} i_{\alpha} \delta_{j_T i_T}$$

$$= N^{\alpha}_i - p_{i_{\underset{\sim}{S}}} \sum_{j_{\underset{\sim}{S}}} \bar{N}^{\alpha}_{i_S j_{\underset{\sim}{S}}} = \sum_{j} N^{\alpha}_j \partial_j d^S_i. \qquad \text{QED}$$

IV.   The Hopf Bifurcation

1.   The Hessian.

Let $X = \Sigma \, X_i \partial_i = \Sigma \, p_i \xi_i \partial_i$ be a vectorfield on $\mathring{\Delta}$. X is a function from $\mathring{\Delta}$ to R^I. We define the Hessian of X at p, $H_p X: T_p \mathring{\Delta} \times T_p \mathring{\Delta} \to R$ to be the bilinear form defined by:

(1.1)
$$H_p X(Y^1, Y^2) = (d_p X(Y^1), Y^2)_p.$$

So to get $H_p X$ we take the derivative of X at p in the Y^1 direction and then take the inner product with Y^2 using the Shahshahani metric at p. If we extend the function X_i and ξ_i to get vectorfields $X = \Sigma \, X_i \partial_i = \Sigma \, x_i \xi_i \partial_i$ on $\mathring{P}$, we can compute:

(1.2)
$$H_x X(Y^1, Y^2) = \sum_{i,j} x_i^{-1} \frac{\partial X_i}{\partial x_j} Y_j^1 Y_j^2$$

$$= \sum_{i} x_i^{-1} \xi_i Y_i^1 Y_i^2 + \sum_{i,j} \frac{\partial \xi_i}{\partial x_j} Y_j^1 Y_j^2.$$

Just as with the corresponding formula for the derivative I.(3.2.4) these formulae taken at $x = p$ are independent of the choice of extending functions provided that the vectors Y^1 and Y^2 lie in the tangent space $T_p \mathring{\Delta} = (R^I)_0 = \{Y \in R^I: \Sigma \, Y_i = 0\}$.

Taking the Hessian at p is itself a linear operation. For X^1 and X^2 vectorfields on $\mathring{\Delta}$ and $t \in R$:

(1.3)
$$H_p(tX^1 + X^2) = t(H_p X^1) + (H_p X^2).$$

In order to study the stability properties of an equilibrium point p of a vectorfield X on $\mathring{\Delta}$, one looks at the derivative

$d_p X: T_p \mathring{\Delta} \to T_p \mathring{\Delta}$ and computes the eigenvalues. The Hessian is important here because:

1 Lemma: With $\{Y^r: r = 1,\ldots,n - 1\}$ an $(\ ,\ )_p$ orthonormal basis for $T_p \mathring{\Delta} = (R^I)_0$, the matrix a_{rs} of the linear map $d_p X: (R^I)_0 \to (R^I)_0$ is given by:

$$(1.4) \qquad a_{rs} = H_p X(Y^s, Y^r) \qquad (r,s = 1,\ldots,n - 1).$$

In particular, the eigenvalues of the linear map $d_p X$ are the same as the eigenvalues of the bilinear form $H_p X$.

Proof: The s column a_{rs} $(r = 1,\ldots,n - 1)$ consists of the coordinates of $d_p X(Y^s)$ with respect to the Y-basis. Since the basis is orthonormal, the Y^r coordinate is obtained by taking the inner product with Y^r. Hence, (1.4). QED

Remark: The eigenvalues of a linear map are independent of the choice of basis. For a bilinear form the independence is only over the choice of orthonormal basis.

Any bilinear form can be decomposed into its symmetric and anti-symmetric (or alternating) parts. So we define;

$$(1.5) \qquad SH_p(Y_1, Y_2) = \frac{1}{2}(H_p(Y_1, Y_2) + H_p(Y_2, Y_1)).$$
$$AH_p(Y_1, Y_2) = \frac{1}{2}(H_p(Y_1, Y_2) - H_p(Y_2, Y_1)).$$

SH_p is symmetric, AH_p is alternating and their sum is H_p, i.e.:

$$SH_p(Y_1,Y_2) = SH_p(Y_2,Y_1).$$

(1.6)
$$AH_p(Y_1,Y_2) = -AH_p(Y_2,Y_1).$$

$$H_p = SH_p + AH_p.$$

This decomposition gives a test for gradient vectorfields.

2 Theorem: A vectorfield on $\mathring{\Delta}$ is a gradient field with respect to the Shahshahani metric iff the Hessian is symmetric at all points of $\mathring{\Delta}$. In detail, $H_pX = SH_pX$ for all p in $\mathring{\Delta}$ or equivalently $AH_pX = 0$ for all p in $\mathring{\Delta}$ iff there exists f: $\mathring{\Delta} \to R$ such that $X(p) = \bar{\nabla}_p f$ for all p in $\mathring{\Delta}$

Proof: The proof is a direct computation. But before diving into it we will describe what is really going on from the tensor analysis point of view.

The vectorfield X is dual with respect to the Shahshahani metric to a differential form ω on the tangent space. In terms of Thm. I.3.1, $\omega_p = X(p)*$. By definition X is the gradient $\bar{\nabla}f$ iff ω is the differential df. So X is a gradient iff ω is an exact form. Now the covariant derivative of ω is a bilinear form and its alternating part is the exterior derivative $d\omega$ (cf. [25, Thm. 5.7]). So the covariant derivative of ω is symmetric iff ω is a closed form. Because $\mathring{\Delta}$ is simply connected closed = exact for linear differential forms. One can actually compute this covariant derivative by using the change of coordinates in Thm. I.4.1. It is not quite the same as the Hessian, essentially because the constant fields ∂_i are not autoparallel with respect to the Shahshahani metric on $\mathring{P}$. However, the two bilinear forms differ only in the symmetric part for any vectorfield X. Putting this all together we get that the alternating part of the Hessian is everywhere zero iff the vector-field is a gradient.

Starting again, suppose X is the gradient of $f: \overset{\circ}{\Delta} \to R$. Extend f to a function on $\overset{\circ}{P} = \{x \in R^I: x_i > 0 \text{ for all } i\}$. For notational convenience define

$$(1.7) \qquad \frac{\overline{\partial f}}{\partial x} = \sum_k x_k \frac{\partial f}{\partial x_k} .$$

By I.(4.12) if $X = \overline{\nabla} f$, then for all i:

$$(1.8) \qquad \xi_i = \frac{\partial f}{\partial x_i} - \frac{\overline{\partial f}}{\partial x} .$$

Taking the partial with respect to x_j we have, using (1.7):

$$(1.9) \qquad \frac{\partial \xi_i}{\partial x_j} = \frac{\partial^2 f}{\partial x_i \partial x_j} - \sum_k x_k \frac{\partial^2 f}{\partial x_k \partial x_j} - \frac{\partial f}{\partial x_j} .$$

Now substitute in (1.2) and note that for $Y^1, Y^2 \in (R^I)_0$ $\Sigma Y_i^1 = 0$ means that the last two terms on the right in (1.9) make no contribution to $H_p(X)$. So we have:

$$(1.10) \qquad H_p(\overline{\nabla} f)(Y^1, Y^2) = \sum_i x_i^{-1} \left(\frac{\partial f}{\partial x_i} - \frac{\overline{\partial f}}{\partial x} \right) Y_i^1 Y_i^2$$

$$+ \sum_{i,j} \frac{\partial^2 f}{\partial x_i \partial x_j} Y_j^1 Y_i^2 \quad (Y^1, Y^2 \in (R^I)_0) .$$

By symmetry of the mixed partial derivatives formula (1.10) shows that $H_p(\overline{\nabla} f)$ is symmetric in Y^1 and Y^2.

For the converse, suppose that X is a vectorfield on $\overset{\circ}{\Delta}$

with $H_p X$ symmetric at every poin p of $\mathring{\Delta}$.   So

(1.11) $$H_x X(Y^1, Y^2) = H_x X(Y^2, Y^1)$$

for $x = p \in \mathring{\Delta}$ and $Y^1, Y^2 \in (R^I)_0$.

In applying (1.2) we can use any extension of X to $\mathring{P}$.   We use the trick introduced in the proof of Prop. III.1.1.   Choose the extension so that each function $\xi_i : \mathring{P} \to R$ is homogeneous of degree -1 i.e. $\xi_i(x) = |x|^{-1} \xi_i(x/|x|)$ with $|x| = \Sigma\, x_i$.   Then by Euler's theorem on homogeneous functions we have for each i:

(1.12) $$\sum_j x_j \frac{\partial \xi_i}{\partial x_j} = -\xi_i.$$

Also we know that $\Sigma\, p_i \xi_i = \Sigma\, X_i = 0$ at every point p of $\mathring{\Delta}$. By homogeneity $\Sigma\, x_i \xi_i = 0$ for all x in $\mathring{P}$.   Taking the partial derivative with respect to x_j we get:

(1.13) $$\sum_i x_i \frac{\partial \xi_i}{\partial x_j} = -\xi_j.$$

I now claim that (1.11) holds for all $x \in \mathring{P}$ and all $Y_1, Y_2 = R^I$. From (1.2) $H_x X(Y^1, Y^2)$ is homogeneous of degree -2 in x and so for $Y^1, Y^2 \in (R^I)_0$ (1.11) holds for all x in $\mathring{P}$ because it holds for $p = x/|x|$ in $\mathring{\Delta}$.   Now since every vector in R^I can be written in the form $Y + tx$ with Y in $(R^I)_0$, the extension of (1.11) to all of R^I follows from:

(1.14) $$H_x X(x, Y^2) = 0 = H_x(Y^1, x) \qquad Y^1, Y^2 \in R^I.$$

This follows from direct substitution in (1.2) using (1.12) when $Y^1 = x$ and (1.13) when $Y^2 = x$.

Thus, the symmetry condition (1.11) holds for all Y^1 and Y^2 in

R^I. Since the first sum in the ξ-version of (1.2) is always symmetric. This symmetry implies that the matrix $(\partial\xi_i/\partial x_j)$ is symmetric, i.e.

$$(1.15) \qquad \frac{\partial\xi_i}{\partial x_j} = \frac{\partial\xi_j}{\partial x_i} \qquad (x \in \overset{\circ}{P},\ i,j \in I).$$

These are the classic integrability conditions for the differential form $\Sigma\ \xi_i dx_i$. By the Poincare Lemma [8, Thm. V.8.1] (1.15) implies that there exists a function $f: \overset{\circ}{P} \to R$ such that

$$(1.16) \qquad \xi_i = \frac{\partial f}{\partial x_i} \qquad (x \in \overset{\circ}{P},\ i \in I).$$

Since $\Sigma\ x_i\xi_i = 0$, $\overline{\frac{\partial f}{\partial x}} = 0$ and so when we restrict f to a function on $\overset{\circ}{\Delta}$, (1.8) implies that X is the gradient $\overline{\nabla}f$. QED

Now we compute the Hessian of our biological vectorfields. Selection is easy since it is a gradient. Apply (1.10):

$$(1.17) \qquad H_p(\overline{\nabla}(\tfrac{1}{2}\ \overline{m}))(Y^1,Y^2) = \sum_i p_i^{-1}(m_i - \overline{m})Y_i^1 Y_i^2 + \sum_{i,j} m_{ij}Y_i^1 Y_j^2$$

$$(Y^1,Y^2 \in (R^I)_0).$$

It is important to see how different the two terms on the right are. Fix $p \in \overset{\circ}{\Delta}$ for the moment and following I.(6.6) write $m_{ij} = \overline{m} + (m_i - \overline{m}) + (m_j - \overline{m}) + \theta_{ij}$. Then since $\Sigma\ Y_i^1 = \Sigma\ Y_j^2 = 0$, (1.17) becomes:

$$(1.18) \qquad H_p(\overline{\nabla}(\tfrac{1}{2}\ \overline{m}))(Y^1,Y^2) = \sum_i p_i^{-1}(m_i - \overline{m})Y_i^1 Y_i^2 + \sum_{i,j} \theta_{ij}Y_i^1 Y_j^2$$

$$(Y^1,Y^2 \in (R^I)_0).$$

So the first term depends on the additive part and the second term depends on the dominance part of the selection matrix m_{ij}.

3 Proposition: Suppose that the recombination and birth rates are completely symmetric so that the recombination field R^S is given by III.2.1. Then in tensor notation the Hessian

$$H_p R^S = (1/4) \sum_{ij} r^S_{ij} b_{ij} d_p(d^S_{ij}) \otimes d_p L^S_{ij} =$$

$$(1/8) \sum_{i,j} r^S_{ij} b_{ij} (p_i p_j + p_{\bar{i}} p_{\bar{j}}) d_p L^S_{ij} \otimes d_p L^S_{ij}$$

$$+$$

(1.19)

$$(1/8) \sum_{i,j} r^S_{ij} b_{ij} d^S_{ij} [d_p(\ln p_i p_j) \otimes d_p \ln p_i p_j - dp(\ln p_{\bar{i}} p_{\bar{j}}) \otimes d_p \ln p_{\bar{i}} p_{\bar{j}}]$$

$$+$$

$$(1/4) \sum_{i,j} r^S_{ij} b_{ij} d^S_{ij} [d_p(\ln p_{\bar{i}} p_{\bar{j}}) \wedge d_p(\ln p_i p_j)]$$

The first two terms are symmetric and so equal $SH_p R^S$. The third term is alternating and so equals $AH_p R^S$.

Proof: We compute the Hessian directly from the definition (1.1). $4R^S = \Sigma \, r^S_{ij} b_{ij} d^S_{ij} \bar{\nabla} L^S_{ij}$. For convenience we will drop the constant factor $r^S_{ij} b_{ij}$ which occurs in all of the sums. It is completely symmetric and so is not affected by the symmetry $i,j \to \bar{i},\bar{j}$. So we assume $4R^S = \Sigma \, d^S_{ij} \bar{\nabla} L^S_{ij}$.

Recall from I.(7.3) that the gradient $\bar{\nabla} L^S_{ij}$ is a constant linear combination of the constant fields ∂_i. So its derivative $d_p(\bar{\nabla} L^S_{ij}) = 0$. Consequently:

$$d_p(4R^S) = \sum d_p(d^S_{ij}) \bar{\nabla}_p L^S_{ij}.$$

$$H_p(4R^S) = \sum d_p(d_{ij}^S) \otimes d_p L_{ij}^S,$$

where we are using the duality between the gradient of L_{ij}^S and its differential.

$$H_p(4R^S) = \sum [p_i p_j d_p(\ln p_i p_j) - p_{\bar{i}} p_{\bar{j}} d_p(\ln p_{\bar{i}} p_{\bar{j}})] \otimes d_p L_{ij}^S.$$

Subtracting and adding the term $p_i p_j d(\ln p_{\bar{i}} p_{\bar{j}})$ in the brackets we break up $H_p(4R^S)$ into two sums. The first $\Sigma_1 =$

$$\sum [p_i p_j d_p(\ln p_i p_j) - p_i p_j d_p(\ln p_{\bar{i}} p_{\bar{j}})] \otimes d_p L_{ij}^S$$

$$= \sum p_i p_j d_p L_{ij}^S \otimes d_p L_{ij}^S$$

$$= \sum p_{\bar{i}} p_{\bar{j}} d_p L_{ij}^S \otimes d_p L_{ij}^S.$$

The last equation holds because the interchange $ij \to \bar{i}\bar{j}$ changes the sign of both $d_p L_{ij}^S$ factors. Averaging the last two sums we see that Σ_1 is 4 times the first term in Prop. 3.

The second sum $\Sigma_2 =$

$$\sum [p_i p_j d_p(\ln p_{\bar{i}} p_{\bar{j}}) - p_{\bar{i}} p_{\bar{j}} d_p(\ln p_{\bar{i}} p_{\bar{j}})] \otimes d_p L_{ij}^S$$

$$= \sum d_{ij}^S d_p(\ln p_{\bar{i}} p_{\bar{j}}) \otimes [d_p(\ln p_i p_j) - d_p(\ln p_{\bar{i}} p_{\bar{j}})].$$

Σ_2 in tern breaks in two. The second sum $\Sigma_{21} =$

$$- \sum d_{ij}^S d_p(\ell_n p_{\bar{i}} p_{\bar{j}}) \otimes d_p(\ln p_{\bar{i}} p_{\bar{j}})$$

$$= \sum d_{ij}^S d_p(\ln p_i p_j) \otimes d_p(\ln p_i p_j).$$

The latter equation is because $ij \to \bar{i}\bar{j}$ changes the sign of d_{ij}^S. Averaging these two we get the second term in Prop. 3.

Finally, $\Sigma_{22} =$

$$\sum d_{ij}^S d_p(\ell n \; p_{\bar{i}}p_{\bar{j}}) \otimes d_p(\ell n \; p_i p_j)$$

$$= -\sum d_{ij}^S d_p(\ell n \; p_i p_j) \otimes d_p(\ell n \; p_{\bar{i}}p_{\bar{j}}).$$

Averaging these two sums we get the final term in Prop. 3 by definition of the wedge product of two forms [24, Sec. 1.9]:

$$(1.20) \qquad \omega_1 \wedge \omega_2 (Y_1, Y_2) = \tfrac{1}{2}[\omega_1(Y_1)\omega_2(Y_2) - \omega_1(Y_2)\omega_2(Y_1)] \qquad \text{QED}$$

Remark: The latter two sums in (1.19) vanish on the Wright manifold $\overset{\circ}{\Lambda}$ since all of the d_{ij}^S's are zero there. So if $p \in \overset{\circ}{\Lambda}$ the Hessian $H_p R^S$ is symmetric and $H_p(-R) = -\Sigma_S H_p(R^S)$ annihilates $T_p\overset{\circ}{\Lambda}$. Furthermore, if R vanishes only on $\overset{\circ}{\Lambda}$, cf. Prop. III.2.6, then $H_p(-R)$ is clearly negative definite on the normal subspace $T_p\overline{\mathfrak{D}}$ to $T_p\overset{\circ}{\Lambda}$.

Finally, for the mutation field $N = \Sigma \; p_j N_{ji}\partial_i$, (1.2) implies:

$$(1.21) \qquad H_p N(Y^1, Y^2) = \sum_{i,j} p_i^{-1} N_{hi} Y_j^1 Y_i^2.$$

This is never symmetric everywhere corresponding to the fact that the mutation field is frequently a gradient field with respect to the usual metric (cf. Thm. III.5.4 and the Remark following) but never is a gradient with respect to the Shahshahani metric.

2. The Wright Conjecture.

At least for the selection plus recombination field, the

Wright Conjecture is essentially true in the zero epistasis cases. We consider these first.

1 Proposition: Consider the two locus, two allele model described in Cor. III.4.2. If the selection field $\bar{\triangledown}(\frac{1}{2}\bar{m})$ has zero epistasis then the combined field $\bar{\triangledown}(\frac{1}{2}\bar{m})$ − R admits a Lyapunov function on $\overset{\bullet}{\Delta}$.

Proof: In this case the sum in III.(4.4) has essentially only one term, i.e. by III.(4.4) and III.(4.13), R is a positive constant times $Q_{14}^{1,b}\bar{\triangledown}(L_{14}^{1,b})^2$. Since $\bar{\triangledown}(\frac{1}{2}\bar{m})$ and R are orthogonal in the zero epistasis case, it follows that for $F = \bar{m} - (L_{14}^{1,b})^2$

$$(\bar{\triangledown}_p(\frac{1}{2}\bar{m}) - R, \bar{\triangledown}_p F)_p = V_A + rQ_{14}^{1,b}\|\bar{\triangledown}_p(L_{14}^{1,b})^2\|_p^2$$

for some positive constant r. This is positive unless both $\bar{\triangledown}_p(\frac{1}{2}\bar{m})$ and $L_{14}^{1,b}(p)$ vanish. QED

For larger models we can prove a local result:

2 Proposition: Suppose: (1) the recombination numbers r_{ij}^s and birth rates b_{ij} are completely symmetric and that the recombination field vanishes only on the Wright manifold (cf. Prop. III.2.6), and (2) fitness is completely symmetric and with zero epistasis so that equation III.(1.12) holds and on each $\overset{\bullet}{\Delta}_\alpha$, $\overline{m}^\alpha$ has a (necessarily unique) non-degenerate critical point q^α.

Then the point $q(i) = \Pi_\alpha q^\alpha(i_\alpha)$ is the unique equilibrium for the combined field $\bar{\triangledown}(\frac{1}{2}\bar{m})$ − R. For every $\epsilon > 0$ sufficiently small, $\bar{m} + \epsilon\hat{H}$ is a Lyapunov function for the combined field on some neighborhood of q.

Proof: Since there is no epistasis, $\bar{\triangledown}(\frac{1}{2}\bar{m})$ and R are orthogonal.

So the combined field vanishes at p iff $R(p) = 0$--and so $p \in \overset{\bullet}{\Lambda}$--and $\bar{\nabla}_p(\frac{1}{2}\bar{m}) = 0$ and so p is a critical point of $\bar{m}$. Since $\bar{m}$ is the sum of the $\overline{m^{\alpha}}$'s, p is a critical point on $\bar{m}$ iff $E^{\alpha}(p)$ is a critical point of $\overline{m^{\alpha}}$ for all α. So as q is the unique point in $\overset{\bullet}{\Lambda}$ with $E^{\alpha}(q) = q^{\alpha}$ for all α, q is the unique $\overset{\bullet}{\Delta}$ equilibrium of the combined field.

   To prove the Lyapunov function result it suffices to show that the function

(2.1) $$f(p) = (\bar{\nabla}_p(\frac{1}{2}\bar{m}) - R, \bar{\nabla}_p(\bar{m} + \epsilon\hat{H}))_p$$

has a nondegenerate local minimum at $p = q$, with $f(q) = 0$. Since $\bar{\nabla}(\bar{m})$ is orthogonal to R, f is the sum of three functions (see III. Secs. 1,2):

$$f_1(p) = (\bar{\nabla}_p(\frac{1}{2}\bar{m}), \bar{\nabla}_p(\bar{m}) = 2\sum_i p_i(m_i - \bar{m})^2$$

(2.2) $$\epsilon f_2(p) = \epsilon(-R, \bar{\nabla}_p\hat{H})_p = \epsilon\sum_{ijs} r^s_{ij} b_{ij} Q^s_{ij} (L^s_{ij})^2$$

$$\epsilon f_3(p) = \epsilon(\bar{\nabla}_p(\frac{1}{2}\bar{m}), \bar{\nabla}_p\hat{H})_p.$$

   At $p = q$: $\bar{\nabla}_p\bar{m}$, R and $\bar{\nabla}_p\hat{H}$ all vanish so it is clear that $f(q) = 0$ and $d_q f = 0$, i.e. q is a critical point. To complete the proof it is enough to show that for $\epsilon > 0$ small enough, the Hessian of the gradient of f is positive definite on $T_q\overset{\bullet}{\Delta} = (R^I)_0$. We will sketch the argument omitting the computational details.

   Apply (1.10) at $p = q$ noting that there $m_i = \bar{m}$ for all i and so the first term in (1.10) vanishes and the second one simplifies to become:

$$(2.3) \quad H_1(Y^1,Y^2) \equiv H_q(\bar{\nabla} f_1)(Y^1,Y^2) = 2 \sum_{ijk} q_i (m_{ij}Y_j^1)(m_{ik}Y_k^2).$$

Now III.(1.12) and the nondegeneracy assumption implies that the annihilator of m_{ij} is exactly $B^{(0)}$ = Kernel of $E: R^I \to \Pi_\alpha R_\alpha$ (see (Cof. II.1.5(d)). Equivalently, the Kernel of the linear map $(R^I)_0 \to R^I$ defined by $Y \to (\Sigma_j m_{ij}Y_j)$ in $B^{(0)}$. This means that the annihilator of H_1 is $T_q\mathcal{D}$ and H_1 is positive definite on the orthogonal complement $T_q\mathring{\Lambda} = T_q\bar{\mathcal{J}}$.

Applying the definition of the Hessian directly, one can show that

$$(2.4) \qquad H_2(Y^1,Y^2) \equiv H_q(\bar{\nabla} f_2)(Y_1,Y_2)$$

$$= \sum_{ijS} r_{ij}^S b_{ij} Q_{ij}^S (\bar{\nabla}_q L_{ij}^S, Y_1)_p (\bar{\nabla}_q L_{ij}^S, Y_2)_p.$$

H_2 clearly annihilates $T_q\mathring{\Lambda}$ and since R vanishes only on Λ_0 the $\bar{\nabla}_q L_{ij}^S$'s with $r_{ij}^S b_{ij} > 0$ span $T_q\mathcal{D}$. So H_2 is positive definite on $T_q\bar{\mathcal{D}}$.

$H_3 \equiv H_2(\bar{\nabla} f_3)$ is somewhat messy to compute. But because of the factor $\bar{\nabla}\bar{m}$ in the definition it is not hard to show that $H_3(Y^1,Y^2) = 0$ if both Y^1 and $Y^2 \in T_q\bar{\mathcal{D}}$. So in particular, $H_2 + H_3$ is the same as H_2 on $T_q\bar{\mathcal{D}}$ and so is positive definite there.

So H_1 is positive definite on $T_q\mathring{\Lambda}$ and annihilates $T_q\bar{\mathcal{D}}$. On the annihilator $T_q\bar{\mathcal{D}}$ $H_2 + H_3$ is positive definite. It then follows from a linear algebra argument that for $\epsilon > 0$ sufficiently small $H_1 + \epsilon(H_2 + H_3)$ is positive definite on all of $T_q\mathring{\Lambda}$. The argument is essentially one used by Smale in an economics context [30]. The precise lemma is stated and proved in [2 Thm. 2.3]. So for $\epsilon > 0$

sufficiently small $H_q(\bar{\nabla} f)$ is positive definite and so f has a non-degenerate local minimum at q.   QED

Remark: There is a spurious argument to show that $\bar{m} + \epsilon\hat{H}$ works on
$\overset{\circ}{\Delta} - G$ where G is any neighborhood of the boundary $\Delta - \overset{\circ}{\Delta}$. It goes
as follows. f_1 is nonnegative on $\overset{\circ}{\Delta}$ vanishing only the fibre
$\bar{\mathfrak{D}}_q = E^{-1}(\{q^\alpha\})$. f_2 is nonnegative on $\overset{\circ}{\Delta}$ vanishing only on $\overset{\circ}{\Lambda}$. f_3
vanishes on $\bar{\mathfrak{D}}_q \cup \overset{\circ}{\Lambda}$ but is otherwise not determined. Recall the
diffeomorphism $E \times L: \overset{\circ}{\Delta} \to \Pi_\alpha \overset{\circ}{\Delta}_\alpha \times R^d$. Now fix G and choose V_0 a
small neighborhood of 0 in R^d. The subset $\Delta: L^{-1}(R^d - V_0) - G$ is
a compact subset of $\overset{\circ}{\Delta}$ disjoint from Λ and so on it f_2 is positive
and bounded away from 0. Since $\bar{\nabla}(\frac{1}{2} \bar{m})$ vanishes on $E^{-1}(\{q^\alpha\})$ there
exists a small neighborhood U_0 of $\{q^\alpha\}$ in $\Pi_\alpha \overset{\circ}{\Delta}_\alpha^\alpha$ suth that on
$E^{-1}(U_0) \cap (L^{-1}(R^d - V_0) - G) = (E \times L)^{-1}(U_0 \times (R^d - V_0)) - G:$
$f_2 > |f_3|$. So on this set $f = f_1 + \epsilon(f_2 + f_3) > 0$ for all $\epsilon > 0$. Now
$E^{-1}((\Pi_\alpha \overset{\circ}{\Delta}_\alpha) - U_0) - G$ is a compact subset of $\overset{\circ}{\Delta}$ disjoint from $\bar{\mathfrak{D}}_q$ and
so on if f_1 is bounded below and $|f_3|$ is bounded above. So for ϵ
small enough $f_1 > \epsilon|f_3|$ on this set. So on this set, too $f > 0$. In
short for an arbitrarily small neighborhood $(E \times L)^{-1}(U_0 \times V_0)$ of q
and $0 < \epsilon < \epsilon_0$ depending on $U_0 \times V_0$ and G, $f > 0$ on
$\overset{\circ}{\Delta} - (E \times L)^{-1}(U_0 \times V_0) - G$. By the above Prop. we can choose $\epsilon_1 > 0$
so that with $0 < \epsilon < \epsilon_1$ f is positive on some punctured neighborhood
of q. The problem is with the order of choices. Fix $\epsilon < \epsilon_1$ and get
U_0, V_0 so that f is positive on $(E \times L)^{-1}(U_0 \times V_0) - \{q\}$. The
ϵ_0 needed to get f positive on the complement of $(E \times L)^{-1}(U_0 \times V_0)$
might be smaller than the ϵ that was fixed to get $U_0 \times V_0$. The best
we can do with all this is to say that ϵ can be chosen small enough
so that the open set $\{p: f(p) < 0\}$ consists of pieces either very

close to $\Delta - \overset{\circ}{\Delta}$ or to q and furthermore that its closure is disjoint from $\overset{\circ}{\Lambda} \cup \overline{\mathfrak{D}}_q$. A sharper argument can probably show that the piece near q is in fact empty and so $\overline{m} + \varepsilon\hat{H}$ works away from the boundary of Δ.

We now turn to the main theorem of the chapter.

<u>3 Theorem</u>: Let X be a smooth vectorfield on $\overset{\circ}{\Delta}$ which is not a gradient field with respect to the Shashahani metric. There exists a smooth one-parameter family of symmetric matrices m^{λ}_{ij} (λ in some neighborhood of 0 in R) such that at $\lambda = 0$ a Hopf bifurcation occurs in the family of vectorfields $\overline{\nabla}(\frac{1}{2} \overline{m^{\lambda}}) + X$. In detail, there exists a point q of $\overset{\circ}{\Delta}$ such that

(a) $\overline{\nabla}_q(\frac{1}{2} \overline{m^{\lambda}}) + X(q) = 0$ for all λ. So q is an equilibrium point for every vectorfield in the family.

(b) With respect to $(,)_q$ the Hessian $H_q(\overline{\nabla}(\frac{1}{2} \overline{m^{\lambda}}) + X)$ has eigenvalues with negative real parts for $\lambda < 0$ and as λ crosses 0 exactly one pair of complex conjugate eigenvalues (with nonzero imaginary part) cross the imaginary axis. if $\rho(\lambda)$ is the real part of this eigenvalue pair then

$$\frac{d\rho(\lambda)}{d\lambda} > 0 \qquad at \qquad \lambda = 0.$$

<u>Proof</u>: Since X is not a gradient, Thm. 1.2 implies that there exists a point $q \in \overset{\circ}{\Delta}$ such that $AH_q(X) \neq 0$. Choose such a point and fix it. q will be the equilibrium point.

We now construct m^{λ}_{ij} in pieces following I.(6.6) by choosing $\overline{m^{\lambda}}$ at q, $m^{\lambda}_i - \overline{m^{\lambda}}$ at q and then θ^{λ}_{ij}.

$\overline{m^{\lambda}}$ at q is arbitrary, choose it to be 0.

$$m_i^\lambda - \overline{m^\lambda} = m_i^\lambda \text{ at } q \text{ is determined by the condition that}$$

$$\overline{\nabla}_q(\tfrac{1}{2}\,\overline{m^\lambda}) + X(q) = 0. \quad \text{So define}$$

$$(2.5) \qquad\qquad k_i = m_i^\lambda - \overline{m^\lambda} = m_i^\lambda = -X_i(q)/q_i$$

So this part will not depend on λ. Define

$$(2.6) \qquad H^a(Y^1,Y^2) = \sum_i q_i^{-1}k_i Y_i^1 Y_i^2 \qquad (Y^1,Y^2 \in (R^I)_0).$$

By (1.18) the symmetric part of the Hessian at q of $\overline{\nabla}(\tfrac{1}{2}\,\overline{m^\lambda}) + X$ consists of $SH_q(X) + H^a$ plus another term depending only on the choice of θ_{ij}^λ. Now I claim that for any symmetric bilinear form H^λ on $(R^I)_0$ there exists a unique choice of θ_{ij}^λ such that

$$(2.7) \qquad H^\lambda(Y^1,Y^2) = \sum_{ij} \theta_{ij}^\lambda Y_i^1 Y_j^2 \qquad (Y^1,Y^2 \in (R^I)_0).$$

The condition that says that θ_{ij}^λ is the pure dominance term at q is

$$(2.8) \qquad\qquad \theta_i^\lambda(q) = \sum_j q_j \theta_{ij}^\lambda = 0 \qquad (i \in I).$$

(2.8) says that, regarded as a symmetric bilinear form on R^I, the matrix θ_{ij}^λ annihilates q. Since $R^I = (R^I)_0 \oplus \{tq: t \in R\}$, we can uniquely extend the $(R^I)_0$ form H^λ to a symmetric bilinear form on R^I by defining:

$$(2.9) \qquad\qquad H^\lambda(q,Y) = H^\lambda(Y,q) = 0 \qquad (Y \in R^I).$$

So given any H^λ on $(R^I)_0$ extend it to R^I by (2.9) and bilinearith. The associated symmetric matrix θ_{ij}^λ of the extended form satisfies (2.8) from (2.9) and (2.7) by definition of the matrix of a bilinear form.

We are left with choosing a one-parameter family H^λ of symmetric bilinear forms on $(R^I)_0$. When they are chosen then the Hessian at q of the combined field becomes:

(2.10)
$$H^\lambda + H^a + SH_q(X) + AH_q(X).$$

So the alternating part $AH_q(X)$ is fixed and the symmetric part $H^\lambda + H^a + SH_q(X)$ is arbitrary. So we can choose it to add constant negative real parts to all of the eigenvalues of $AH_q(X)$ except for one imaginary pair and there let the real part that is added on be λ.

In detail, define the linear map $L: (R^I)_0 \to (R^I)_0$ by

(2.11)
$$(L(Y^1), Y^2)_q = AH_q(Y^1, Y^2).$$

With respect to the inner product $(\ ,\ )_q$, L is a skew-symmetric operator. So we can choose an $(\ ,\ )_q$ orthonormal basis $Y^1, \ldots, Y^{n-1}$ such that for real numbers $t_1, \ldots t_k$, $0 < k \leq (n-1)/2$:

$$L(Y^1) = t_1 Y^2 \qquad L(Y^2) = -t_1 Y^1$$

$$L(Y^3) = t_2 Y^4 \qquad L(Y^4) = -t_2 Y^3$$

(2.12)
$$\vdots$$

$$L(Y^{2k-1}) = t_k Y^{2k} \qquad L(Y^{2k}) = -t_k Y^{2k-1}$$

$$L(Y^i) = 0 \qquad\qquad 2k < i \leq n-1.$$

Now define $S^\lambda: (R^I)_0 \to (R^I)_0$ by

$$S^\lambda(Y^1) = \lambda Y^1, \quad S^\lambda(Y^2) = \lambda Y^2$$

(2.13)
$$S^\lambda(Y^i) = -Y^i \quad 2 < i \leq n-1.$$

The associated symmetric form is defined by:

(2.14)
$$H_0^\lambda(Y^1, Y^2) = (S^\lambda(Y^1), Y^2)_q.$$

Finally, define

(2.15)
$$H^\lambda = H_0^\lambda - H^a - SH_q(X).$$

This defines θ_{ij}^λ and hence m_{ij}^λ with Hessian at q equal to (cf. (2.10)):

(2.16)
$$H_0^\lambda + AH_q(X) \qquad \text{QED}$$

We now apply the bifurcation theorem of Hopf [22, Thm. 3.15] and get

<u>4 Corollary</u>: Let X be a smooth vectorfield on $\overset{\circ}{\Delta}$ which is not a gradient field with respect to the Shahshahani metric (e.g. the recombination field R or the mutation field N) then there exists a symmetric matrix m_{ij} of fitness constants such that the combined field $\bar{\nabla}(\frac{1}{2}\bar{m}) + X$ has a nontrivial (i.e. nonequilibrium) periodic orbit.

The theorem and its corollary leave open some important questions:

What is the role of position effects? That is, can the m_{ij}'s be chosen completely symmetric? For the proof we needed the power to construct any symmetric form H^λ and this required us to range over all symmetric matrices m_{ij}.

What about the stability of the cycles? Can attracting, i.e. limit, cycles occur? In any case are they structurally stable?

We can't answer questions like these by the pure thought arguments of the above proof. The construction and detailed analysis of examples is needed. For example, if examples can be constructed satisfying the "vague attractor" condition of [21] then both of the latter questions would have affirmative answers. The place to go for examples is the simplest case. These cycles can occur in the two allele model--by the above corollary. Investigation of this case is in progress.

Appendix

1. Proper Mappings.

Lemma II.1.2 is a special case of the following:

Theorem: Let f: X → Y be a topologically proper local homeomorphism of locally 1-connected Hausdorff spaces (1-connected means connected and simply connected). f is a finite covering space. In particular, if X is connected and Y is 1-connected then f is a homeomorphism.

This in turn follows from:

Lemma: Let f: X → Y be a topologically proper local homeomorphism of Hausdorff spaces. f has the unique homotopy lifting property: Let Z be a compact, locally connected Hausdorff space and assume that we have a commutative diagram of continuous maps:

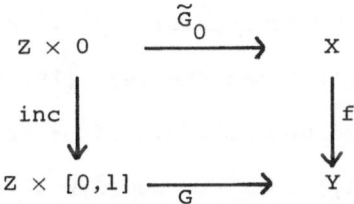

G lifts uniquely to a continuous map $\widetilde{G}$: Z × [0,1] → X with $\widetilde{G}_0$ the given map and $f \cdot \widetilde{G} = G$.

Proof: (1) Because f is a local homeomorphism of Hausdorff spaces, path lifts, when they exist, are unique.

(2) For y ∈ Y, $f^{-1}y$ is compact because f is proper and it is discrete because f is a local homeomorphism. So $f^{-1}y$ is finite.

(3) Define $0 \leq t^* \leq 1$ to be the supremum of the set

{s: G|Z × [0,s] lifts continuously to $\tilde{G}$ with $\tilde{G}_0$ the given map}.

(4) We show that the supremum is achieved. This is clear if t* = 0. If t* > 0 then by (1) $\tilde{G}$ can be defined uniquely to lift G on Z × [0,t*) where the latter factor is open on the right. Now $f^{-1}G(Z × [0,1])$ is compact and so for each z ∈ Z the limit set

$$S_z = \cap_{0<t<t*} \text{Closure}[\tilde{G}(z × [t,t*))]$$

is the decreasing intersection of nonempty compact connected sets and so is nonempty, compact and connected (Kelley, [19], p. 163). Furthermore, if x ∈ S_z then there is a sequence t_α converging up to t* such that x = Lim $\tilde{G}(z,t_\alpha)$ (Kelley, [19], p. 72). So f(x) = Lim G(z,t_α) = G(z,t*). Thus, S_z is a nonempty connected subset of the finite set $f^{-1}G(z,t*)$. So S_z consists of a single point which we will denote $\tilde{G}(z,t*)$. We must show that this extension is continuous. Let U and V be neighborhoods of $\tilde{G}(z,t*)$ and G(z,t*) respectively such that f: U → V is a homeomorphism. $G^{-1}V$ is a neighborhood of (z,t*) and so by Wallace's Lemma (Kelley, [19, p. 142]) there exists ε > 0 and W a connected neighborhood of z such that W × [t* − ε,min(t* + ε,1)) ⊂ $G^{-1}V$. Now $\tilde{G}$|W × [t* − ε,t*) and $(f|U)^{-1}$·G|W × [t* − ε,t*) agree on z × [t* − ε,t*) and so by connectedness they are equal. The latter map extends continuously to W × [t* − ε,min(t* + ε,1)] and so gives a continuous extension of the former.

(5) We have just shown not only that $\tilde{G}$ extends to Z × [0,t*] but also that it extends to an open subset of Z × [0,1] containing Z × [0,t*], namely to the union of Z × [0,t*] and the union of the family {W × [t* − ε,min(t* + ε,1))} indexed by z ∈ Z. If

t* < 1, Wallace's Lemma again and compactness of Z imply that the open set contains some Z × [0,t**] with t* < t** ≤ 1. This contradicts the definition of t*. So t* = 1.   QED

Remark: Note that if $Z_0 \subseteq Z$ and we began with a lifting defined on Z × 0 ∪ Z_0 × [0,1] our lift $\tilde{G}$ agrees with the given lift on Z_0 × [C,1] by uniqueness.

Proof of Theorem: (See Spanier, [31, p. 78].) Let V be a 1-connected open neighborhood of $y_0 \in Y_0$. Let $\{x_1,\ldots,x_n\} = f^{-1}y_0$ and let U_i be the arc-component of x_i in $f^{-1}V$. We show that $U_i \cap U_j = \emptyset$ if $i \neq j$ and f: $U_i \to V$ is bijective for all i.

(1) Any point y_1 of V can be connected by a path y_t in V, to y_0. There are n lifts of this path beginning at $x_1,\ldots,x_n$ respectively. Thus, every point of $f^{-1}V$ connects in $f^{-1}V$ to a point of $f^{-1}y_0$. So $f^{-1}V = \cup_i U_i$ and $fU_i = V$ for each i.

(2) If a point of $f^{-1}V$ connects in $f^{-1}V$ to both x_i and x_j (i.e. $U_i \cap U_j \neq \emptyset$), then x_i and x_j can be connected by a path in $f^{-1}V$. This path projects to a loop in V which is homotopic in V to the constant path at y. By the Lemma this homotopy lifts to a homotopy of the original path to a path connecting x_i and x_j in $f^{-1}y$ which is discrete. So i = j. Contrapositively, $U_i \cap U_j = \emptyset$ if $i \neq j$.

(3) The argument of (2) also shows that $f|U_i$ is injective for each i, because no two points in the same fibre can be connected by a path in $f^{-1}V$.   QED

2. Partially Defined Distributions.

The conjecture remarked upon after Thm. II,2.7 can be written as:

Property I(K): For a complex K, every compatible family of distributions on the subproducts I_S corresponding to the blocs S in K is induced by some distribution on I i.e.

(2.1)
$$E^K(\Delta) = V_K \cap \Pi\{\Delta_S : S \in K\}.$$

A related conjecture is:

Property H(K): For a complex K, if a family of interior distributions $\{p^S : S \in K\}$ $(\in \Pi\{\mathring{\Delta}_S : S \in K\})$ is induced by some distribution p on I, i.e. $E^K(p) = \{p^S\}$, then the distribution of maximum entropy π with $E^K(\pi) = \{p^S\}$ is an interior distribution, i.e. $\pi \in \mathring{\Delta}$

For our examples, let $\ell = 3$ and $I_\alpha = \{0,1\}$ for $\alpha = 1,2,3$. The points of I are the vertices of the unit cube in R^3. Let K consist of all subsets of $L = \{1,2,3\}$ except for L itself. So we are given compatible families of pairwise distributions and are looking for a distribution on the product.

For our first example, let p_0 put weight 1/6 on each of the vertices except for the diagonal pair $(0,0,0)$ and $(1,1,1)$ and weight zero on these. $p_0 \in \Delta - \mathring{\Delta}$ but projects to a family of interior distributions. In fact, the projection to each face of the cube puts on weights: 1/3, 1/3, 1/6 and 1/6. To show that p_0 is the member of Δ with maximum entropy among those with the specified projections, we will show that it is the only such member of Δ.

In this case the Kernel of E^K is one dimensional consisting of all multiples of the vector x satisfying

(2.2) $\qquad x(i_1, i_2, i_3) = (-1)^\sigma$ where $\sigma = i_1 + i_2 + i_3$.

So for any vector in the kernel, as one steps from one vertex to another along an edge the value of x just changes sign. In particular, $x(0,0,0)$ and $x(1,1,1)$ have opposite signs. Hence $p_0 + tx \in \Delta$ iff $t = 0$.

For our second example, we use the leverage indicated by the first. For $\varepsilon > 0$, let x_ε take the value $(1 + \varepsilon)/6$ on the six points other than $(0,0,0)$ and $(1,1,1)$ and let x_ε be $-\varepsilon/2$ on each of these elements. For ε small--in fact, for $\varepsilon < 1/2$--x_ε projects under E^K to an element of $\Pi(\mathring{\Delta}_S)$. But for all t $x_\varepsilon + tx$ is negative on either $(0,0,0)$ or $(1,1,1)$. So no member of Δ maps to $E^K(x_\varepsilon) \in V_K \cap \Pi(\mathring{\Delta}_S)$.

To be specific, with $\varepsilon = 1/4$, consider the distribution on $\{0,1\} \times \{0,1\}$:

$$p(0,0) = p(1,1) = 1/12$$
(2.3)
$$p(0,1) = p(1,0) = 5/12.$$

Putting this distribution on each of the pairwise subproducts of $I = \{0,1\}^3$ yields a compatible family of distributions which is not induced by any distribution on I.

Proposition: For any complex K and any family of distributions $\{p^S : S \in K\}$ in the image $E^K(\Delta)$, let π be the distribution of maximum entropy mapping to $\{p^S\}$ under E^K. $\{p^S\} \in \Delta_K - \mathring{\Delta}_K$, i.e. the entire preimage $(E^K|\Delta)^{-1}\{p^S\} \subset \Delta - \mathring{\Delta}$ iff $\pi \in \Delta - \mathring{\Delta}$.

Proof: If $E^K(p_0) = E^K(p_1) = \{p^S\}$. Then the whole segment $p_t = p_0 + t(p_1 - p_0)$ maps under E^K to $\{p^S\}$. Suppose $p_0 \in \Delta - \mathring{\Delta}$ and $p_1 \in \mathring{\Delta}$. I claim that for $t > 0$ and small enough $H(p_t) > H(p_0)$ where H is the entropy. So p_0 cannot be π. Thus, if there is any interior distribution in the fibre $(E^K|\Delta)^{-1}\{p^S\}$ then π is interior.

The function $H(p_t)$ is strictly increasing near $t = 0$ because as a function of t it is differentiable for $t > 0$ and the derivative approaches $+\infty$ as t approaches 0. This is because the derivative of $-t \ln t$ approaches $+\infty$ as t approaches 0. QED

Actually the above result can be proved directly from Thm. 11.1.6.

Corollary: For any complex K, the property $H(K)$ implies $I(K)$.

Proof: If (2.1) is false, i.e. $E^K(\Delta) = \Delta_K$ is a proper subset of $V_K \cap \Pi\{\Delta_S\}$ then there must exist $\{p_0^S\} \in V_K \cap \Pi\{\mathring{\Delta}_S\} - \Delta_K$. Let $\{p_1^S\} \in \mathring{\Delta}_K = E^K(\mathring{\Delta})$. The segment between them lies in $V_K \cap \Pi\{\mathring{\Delta}_S\}$ and must meet some point $\{p_t^S\}$ of $E^K(\Delta) - E^K(\mathring{\Delta})$. By the theorem, the maximum entropy distribution π mapping to $\{p_t^S\}$ must lie in $\Delta - \mathring{\Delta}$. This contradicts $H(K)$. Taking the contrapositive, we get the corollary. QED

3.   Game Dynamics.

In a recent, elegant paper [33] Taylor and Jonker give a dynamic interpretation of the concept due to Maynard Smith and Price of an evolutionarily stable strategy in a biological game.   Their dynamic model turns out to be identical to the vectorfield model of frequency dependent selection.   Using the concept of the Hessian from Chap. IV we get a more conceptual proof of their main result.

In this case I is the set of n strategies and p_i is the proportion of the population using strategy i. $F(i|p)$ denotes the payoff to a player using strategy i when the strategy distribution vector of the population is p.   Taylor and Jonker examine the differential equation:

(3.1)
$$\frac{dp_i}{dt} = p_i(F(i|p) - F(p|p)).$$

Here $F(p|p) = \Sigma\ p_i F(i|p)$.   So defining $\xi_i(p) = F(i|p) - F(p|p)$ we see that this equation comes from the vectorfield on Δ:

$$X(p) = \sum P_i \xi_i(p)\partial_i.$$

Now define:

(3.2)
$$a_{ij} = \frac{\partial F(i|p)}{\partial p_j}$$

where this really means extend the functions $F(i|p)$ to functions $F(i|x)$ on R^I and let $a_{ij} = \partial F(i|x)/\partial x_j$.

An equilibrium $p \in \overset{\circ}{\Delta}$, i.e. $F(i|p) = F(p|p)$ for all i, is called an evolutionarily stable equilibrium or ESS if

(3.3) $$\sum a_{ij} Y_i Y_j < 0 \qquad (0 \neq Y \in (R^I)_0).$$

Since $\xi_i(p) = 0$ for all i, it is easy to check from IV.(1.2), that the Hessian of X at p is given by:

(3.4) $$H_p(Y^1, Y^2) = \sum a_{ij} Y_j^1 Y_i^2 \qquad (Y^1, Y^2 \in (R^I)_0).$$

So p is an ESS iff $H_p(Y,Y) < 0$ for $Y \neq 0$ in $(R^I)_0$.

We now apply the following classical theorem of Lyapunov. It is a matrix theory result with a simple differential equation proof.

Lemma: Let V be a Euclidean vector space with inner product $(,)$. Let $A: V \to V$ be a linear map with associated bilinear form:

(3.5) $$H(Y^1, Y^2) = (AY^1, Y^2).$$

With the symmetrized form $SH(Y^1, Y^2) = \frac{1}{2}(H(Y^1, Y^2) + H(Y^2, Y^1))$ is associated the self-adjoint linear map $A_S: V \to V$ by

(3.6) $$SH(Y^1, Y^2) = (A_S Y^1, Y^2).$$

A_S has only negative real eigenvalues iff SH is negative definite, i.e. $SH(Y,Y) < 0$ if $Y \neq 0$, and in that case the eigenvalues of A have negative real parts.

Proof: Recall that with respect to an orthonormal basis the matrix of the bilinear form and the associated linear map are the same (see Lemma IV.1.1). Since a self-adjoint operator can be diagonalized with respect to some orthonormal basis, it easily follows that SH is negative definite iff A_S has only negative eigenvalues. If this is true then define the differential equation on V:

$$\frac{dY}{dt} = AY.$$

Let $F(Y) = -\frac{1}{2}(Y,Y)$. Then

$$\frac{dF}{dt} = -(AY,Y) = -H(Y,Y) = -SH(Y,Y) > 0 \qquad (Y \neq 0).$$

So $F(Y)$ is a Lyapunov function for Y and every solution of the differential equation approaches 0 asymptotically. This implies that the eigenvalues of A have negative real parts. For a real eigenvalue a gives a solution of the form $e^{at}Y_0$. A complex conjugate pair $a \pm ib$ gives solutions of the form $e^{at}[\cos bt\, Y_1 + \sin bt\, Y_2]$. These only approach 0 as t goes to ∞ if $a < 0$. \hfill QED

Theorem: Let p be an ESS in $\overset{\circ}{\Delta}$. Then p is an asymptotically stable equilibrium.

Proof: Since p is an ESS the symmetrized Hessian $SH_p(X)$ is negative definite. So by the above lemma and Lemma IV.1.1 the linearization of the vectorfield $d_pX: T_p\overset{\circ}{\Delta} \to T_p\overset{\circ}{\Delta}$ has eigenvalues with negative real parts. So p is asymptotically stable. \hfill QED

Jonker and Taylor also prove the above theorem when the equilibrium lies on the boundary $\Delta - \overset{\circ}{\Delta}$, with an adjustment of the definition of an ESS. As we have not dealt with problems at the boundary earlier we won't pursue in that direction.

The payoff function $F(i|p)$ is a generalization of the payoff matrix a_{ij} = payoff i receives playing against j. In that case $F(i|p) = \Sigma\, p_i a_{ij} \equiv a_{ip}$. For the linear game the vectorfield is

$X = \Sigma \, p_i(a_{ip} - a_{pp})$. This is just like our selection field except that a_{ij} need not be symmetric. If a_{ij} is symmetric a game theorist would call the game completely cooperative. Each player benefits as much as his "opponent" and so the population cooperatively evolves to the point of maximum mean payoff a_{pp}. This is because in the symmetric case the vectorfield is the gradient $\bar{\nabla}(\frac{1}{2} a_{pp})$. In fact, this case is just a notational change from the selection field. I like thinking of this as saying that sex is a cooperative game.

This suggests the general notion of a cooperative game is one where the vectorfield associated to the payoff function $F(i|p)$ is a gradient with respect to the Shahshahani metric. Thm. IV.1.2 provides a test for when this is true namely when the Hessian is everywhere symmetric.

At the other extreme if a_{ij} is anti-symmetric i.e. $a_{ij} = -a_{ji}$ then the game theorist calls the game zero sum because the joint pay-off in an i vs j game is $a_{ij} + a_{ji} = 0$. Since every matrix can be written as the sum of a symmetric and an anti-symmetric matrix, every linear game equation is the sum of a gradient and a zero-sum game field.

It would be interesting to see if the dynamical systems associated to zero-sum games have special properties.

Bibliography

[1]   Abraham, R. _Introduction to Morphology_ (preprint).

[2]   Akin, E. "On Pareto Optimality" (to appear).

[3]   Brocker, Th. _Differentiable Germs and Catastrophes_, London
        Math. Soc. Lect. Notes # 17, Cambridge U.P. (1975).

[4]   Cockerham, C.C. "An Extension of the Concept of Partitioning
        Hereditary Variance for Analysis of Covariance among Relatives
        when Epistasis is Present" (1954) Genetics, 39: 859-882.

[5]   Cole, L.C. "The Population Consequences of Life History
        Phenomena" (1954) Q. Rev. Biol. 29: 103-137.

[6]   Crow, J.F. and Kimura, M. _An Introduction to Population
        Genetics Theory_, Harper and Row Publishers, Inc. (1970).

[7]   Demetrius, L. "Primitivity Conditions for Growth Matrices"
        (1971) Math. Biosciences 12: 53-58.

[8]   Edwards Jr., C.H. _Advanced Calculus of Several Variables_,
        Academic Press Inc. (1973).

[9]   Ewens, W.J. "With Additive Fitness, the Mean Fitness Increases"
        (1969) Nature 221: 1076.

[10]   Ewens, W.J. _Population Genetics_, Metheun and Co., Ltd (1969).

[11]   Felsenstein, J. "The Effect of Linkage on Directional Selec-
        tion" (1965) Genetics 52: 349-363.

[12]   Gokhale, D.V. and Kullback, S. _The Information in Contingency
        Tables_, Marcel Dekker, Inc. (1978).

[13]   Halmos, P.R. _Finite Dimensional Vector Spaces_ (2nd edition)
        D. Van Nostrand Company, Inc. (1958).

[14]   Hirsch, M., Pugh, C. and Shub, M. _Invariant Manifolds_, Lecture
        Notes in Math # 583, Springer Verlag (1977).

[15]   Hirsch, M. and Smale, S. _Differential Equations, Dynamical
        Systems, and Linear Algebra_, Academic Press, Inc. (1974).

[16]   Hoppensteadt, F. _Mathematical Theories of Populations: Demo-
        graphics, Genetics and Epidemics_, SIAM (1975).

[17]   Jacquard, A. _The Genetic Structure of Populations_, Springer-
        Verlag (1974).

[18]   Karlin, S. _A First Course in Stochastic Processes_, Academic

Press, Inc. (1966).

[19]   Kelley, J.L. *General Topology*, D. Van Nostrand Company, Inc. (1955).

[20]   Kempthorne, O. *An Introduction to Genetic Statistics*, The Iowa State U.P. (1969).

[21]   Kullback, S. *Information Theory and Statistics,* John Wiley and Sons, Inc. (1959).

[22]   Marsden, J.E. and McCracken, M. *The Hopf Bifurcation and its Applications*, Springer-Verlag (1976).

[23]   May, R.M. *Stability and Complexity in Model Ecosystems*, Princeton U.P. (1973).

[24]   Moran, P.A.P. "On the Nonexistence of Adaptive Topographies" (1964) Ann. Human Genet. 27: 383-393.

[25]   Nelson, E. *Tensor Analysis*, Princeton U P. (1967).

[26]   O'Neill, B. "The Fundamental Equations of a Submersion" (1966) Michigan Math. Jour., 13: 459-469.

[27]   Palais, R.S. *Foundations of Global Nonlinear Analysis*, W.A. Benjamin, Inc. (1968).

[28]   Shahshahani, S. *A New Mathematical Framework for the Study of Linkage and Selection*, Memoirs AMS. # 211 (1979).

[29]   Smale, S. "Sufficient Conditions for an Optimum" in *Dynamical Systems--Warwick 1974*, Lect. Notes in Math. # 468 Springer-Verlag (1975).

[30]   Smale, S. "On the Differential Equations of Populations in Competition" (1976) J. Math. Bio. 3: 5-7.

[31]   Spanier, E.H. *Algebraic Topology*, McGraw-Hill, Inc. (1966).

[32]   Spivak, M. *Calculus on Manifolds*, W.A. Benjamin, Inc. (1965).

[33]   Taylor, P.D. and Jonker, L.B. "Evolutionarily Stable Strategies and Game Dynamics" (1978) Math. Biosciences 40: 145-156.

[34]   Thom, R. *Structural Stability and Morphogenes*, W.A. Benjamin, Inc. (1975).

[35]   Wright, S. "Adaptation and Selection" in *Genetics, Paleontology and Evolution*, Ed. by Jepsen, Simpson and Mayr, Princeton U. P. (1949).

[36]   Wright, S. "'Surfaces' of Selective Value" (1967) PNAS 58: 165-172.

[37] Wright, S. "Random Drift and the Shifting Balance Theory of Evolution" in <u>Mathematical Topics in Population Genetics</u>, Ed. by Kojima, Springer-Verlag (1970).

Index

Bio— mathematics

Managing Editors: K. Krickeberg, S. A. Levin

Editorial Board: H. J. Bremermann, J. Cowan,
W. M. Hirsch, S. Karlin, J. Keller, R. C. Lewontin,
R. M. May, J. Neyman, S. I. Rubinow, M. Schreiber,
L. A. Segel

Volume 1:
Mathematical Topics in Population Genetics
Edited by K. Kojima
1970. 55 figures. IX, 400 pages
ISBN 3-540-05054-X

"...It is far and away the most solid product I have
ever seen labelled biomathematics."
American Scientist

Volume 2: E. Batschelet
Introduction to Mathematics for Life Scientists
2nd edition. 1975. 227 figures. XV, 643 pages
ISBN 3-540-07293-4

"A sincere attempt to relate basic mathematics to the
needs of the student of life sciences."
Mathematics Teacher

M. Iosifescu, P. Tăutu
**Stochastic Processes and Applications in Biology
and Medicine**

Volume 3
Part 1: **Theory**
1973. 331 pages.
ISBN 3-540-06270-X

Volume 4
Part 2: **Models**
1973. 337 pages
ISBN 3-540-06271-8

Distributions Rights for the Socialist Countries:
Romlibri, Bucharest

"... the two-volume set, with its very extensive bibliography, is a survey of recent work as well as a text-
book. It is highly recommended by the reviewer."
American Scientist

Volume 5: A. Jacquard
The Genetic Structure of Populations
Translated by B. Charlesworth, D. Charlesworth
1974. 92 figures. XVIII, 569 pages
ISBN 3-540-06329-3

"...should take its place as a major reference work.."
Science

Volume 6: D. Smith, N. Keyfitz
Mathematical Demography
Selected Papers
1977. 31 figures. XI, 515 pages
ISBN 3-540-07899-1

This collection of readings brings together the major
historical contributions that form the base of current
population mathematics tracing the development of
the field from the early explorations of Graunt and
Halley in the seventeenth century to Lotka and his
successors in the twentieth. The volume includes
55 articles and excerpts with introductory histories
and mathematical notes by the editors.

Volume 7: E. R. Lewis
Network Models in Population Biology
1977. 187 figures. XII, 402 pages
ISBN 3-540-08214-X

Directed toward biologists who are looking for an
introduction to biologically motivated systems
theory, this book provides a simple, heuristic
approach to quantitative and theoretical population
biology.

Springer-Verlag
Berlin
Heidelberg
New York

A
Springer
Journal

Journal of

Mathematical Biology

Ecology and Population Biology
Epidemiology
Immunology
Neurobiology
Physiology
Artificial Intelligence
Developmental Biology
Chemical Kinetics

Edited by H.J. Bremermann, Berkeley, CA; F.A. Dodge,
Yorktown Heights, NY; K.P. Hadeler, Tübingen; S.A. Levin,
Ithaca, NY; D. Varjú, Tübingen.

Advisory Board: M.A. Arbib, Amherst, MA; E. Batschelet, Zürich;
W. Bühler, Mainz; B.D. Coleman, Pittsburgh, PA; K. Dietz,
Tübingen; W. Fleming, Providence, RI; D. Glaser, Berkeley, CA;
N.S. Goel, Binghamton, NY; J.N.R. Grainger, Dublin;
F. Heinmets, Natick, MA; H. Holzer, Freiburg i. Br.; W. Jäger,
Heidelberg; K. Jänich, Regensburg; S. Karlin, Rehovot/Stanford
CA; S. Kauffman, Philadelphia, PA; D.G. Kendall, Cambridge;
N. Keyfitz, Cambridge, MA; B. Khodorov, Moscow; E.R. Lewis,
Berkeley, CA; D. Ludwig, Vancouver; H. Mel, Berkeley, CA;
H. Mohr, Freiburg i. Br.; E.W. Montroll, Rochester, NY; A. Oaten,
Santa Barbara, CA; G.M. Odell, Troy, NY; G. Oster, Berkeley, CA;
A.S. Perelson, Los Alamos, NM; T. Poggio, Tübingen;
K.H. Pribram, Stanford, CA; S.I. Rubinow, New York, NY;
W.v. Seelen, Mainz; L.A. Segel, Rehovot; W. Seyffert, Tübingen;
H. Spekreijse, Amsterdam; R.B. Stein, Edmonton; R. Thom,
Bures-sur-Yvette; Jun-ichi Toyoda, Tokyo; J.J. Tyson, Blacks-
bough, VA; J. Vandermeer, Ann Arbor, MI.

Springer-Verlag
Berlin
Heidelberg
New York

Journal of Mathematical Biology publishes papers in which
mathematics leads to a better understanding of biological phenomena, mathematical papers inspired by biological research and
papers which yield new experimental data bearing on mathematical models. The scope is broad, both mathematically and biologically and extends to relevant interfaces with medicine,
chemistry, physics and sociology. The editors aim to reach an
audience of both mathematicians and biologists.